Lecture Notes in Networks and Systems 907

The series “Lecture Notes in Networks and Systems” publishes the latest developments in Networks and Systems—quickly, informally and with high quality. Original research reported in proceedings and post-proceedings represents the core of LNNS.

Volumes published in LNNS embrace all aspects and subfields of, as well as new challenges in, Networks and Systems.

The series contains proceedings and edited volumes in systems and networks, spanning the areas of Cyber-Physical Systems, Autonomous Systems, Sensor Networks, Control Systems, Energy Systems, Automotive Systems, Biological Systems, Vehicular Networking and Connected Vehicles, Aerospace Systems, Automation, Manufacturing, Smart Grids, Nonlinear Systems, Power Systems, Robotics, Social Systems, Economic Systems and other. Of particular value to both the contributors and the readership are the short publication timeframe and the world-wide distribution and exposure which enable both a wide and rapid dissemination of research output.

The series covers the theory, applications, and perspectives on the state of the art and future developments relevant to systems and networks, decision making, control, complex processes and related areas, as embedded in the fields of interdisciplinary and applied sciences, engineering, computer science, physics, economics, social, and life sciences, as well as the paradigms and methodologies behind them.

Indexed by SCOPUS, INSPEC, WTI Frankfurt eG, zbMATH, SCImago.

All books published in the series are submitted for consideration in Web of Science.

For proposals from Asia please contact Aninda Bose (aninda.bose@springer.com).

Grzegorz Sierpiński · Salah Al-Majeed ·
Elżbieta Macioszek
Editors

Advanced Solutions for Mobility in Urban Areas

Springer

Editors
Grzegorz Sierpiński
Faculty of Transport and Aviation Engineering
Silesian University of Technology
Gliwice, Poland

Elżbieta Macioszek
Faculty of Transport and Aviation Engineering
Silesian University of Technology
Gliwice, Poland

Salah Al-Majeed
Al Akhawayn University
Ifrane, Morocco

ISSN 2367-3370 ISSN 2367-3389 (electronic)
Lecture Notes in Networks and Systems
ISBN 978-3-031-53180-4 ISBN 978-3-031-53181-1 (eBook)
https://doi.org/10.1007/978-3-031-53181-1

This Springer imprint is published by the registered company Springer Nature Switzerland AG
The registered company address is: Gewerbestrasse 11, 6330 Cham, Switzerland

Paper in this product is recyclable.

Preface

The mobility of the future is one of the challenges both in terms of the development of transport infrastructure, transport organization, and new ways of traveling. The aim is to achieve clean and efficient transport. Among the solutions, an important place is the need to improve urban public transport and the expansion of infrastructure encouraging the choice of walking and cycling. We should also remember about people with special needs and solutions dedicated to them.

This publication contains selected papers submitted to and presented at the 19th "Transport Systems. Theory and Practice" Scientific and Technical Conference organized by the Department of Transport Systems, Traffic Engineering and Logistics in the Faculty of Transport and Aviation Engineering at the Silesian University of Technology (Katowice, Poland). The problems addressed in the publication entitled *Advanced Solutions for Mobility in Urban Areas* include results of case study-based scientific and practical approaches to the problems of contemporary transport systems.

The authors drew attention to challenges for public transport organizers including effectiveness, schedule planning problems, integration in public transport, or different factors which have influence on travel choices. For all these situations, it is necessary to apply solutions that take into account the needs in the field of environmental protection, as well as the reduction of energy consumption. The publication also includes an abundance of numerical analyses based on significant datasets, illustrating the close affiliation between mobility and quality of life.

We would like to express our deepest gratitude to all authors, for reflecting the key problems of contemporary transport systems in a concise manner, as well as to reviewers, in recognition of their insightful remarks and suggestions without which this collection of papers would have never been published.

September 2023

Grzegorz Sierpiński
Salah Al-Majeed
Elżbieta Macioszek

Organization

The 19th Scientific and Technical Conference "Transport Systems. Theory and Practice" (TSTP2023) is organized by the Department of Transport Systems, Traffic Engineering and Logistics, Faculty of Transport and Aviation Engineering, Silesian University of Technology (Poland).

Organizing Committee

Organizing Chair

Grzegorz Sierpiński	Silesian University of Technology, Poland

Members

Agnieszka Popowicz
Adrian Barchański
Grzegorz Karoń
Marcin Kłos
Krystian Szewczyński
Aleksander Sobota
Piotr Soczówka
Agata Wyderka
Marcin Staniek
Ireneusz Celiński

The Conference Took Place Under the Honorary Patronage

Ministry of Infrastructure
Silesian Voivode
Marshal of the Silesian Voivodeship

Scientific Committee

Co-chairs

Grzegorz Sierpiński	Silesian University of Technology, Poland
Elżbieta Macioszek	Silesian University of Technology, Poland

Rahmi Akçelik	SIDRA SOLUTIONS, Australia
Werner Brilon	Ruhr-University Bochum, Germany
Jacek Chmielewski	Cracow University of Technology, Poland
Margarida Coelho	University of Aveiro, Portugal
Csaba Csiszár	Budapest University of Technology and Economics, Hungary
Roberta Di Pace	University of Salerno, Fisciano, Italy
Mehmet Dikmen	Baskent University, Turkey
Domokos Esztergár-Kiss	Budapest University of Technology and Economics, Hungary
Zoltán Fazekas	Institute for Computer Science and Control, Hungary
Piotr Folęga	Silesian University of Technology, Poland
József Gál	University of Szeged, Hungary
Anna Granà	University of Palermo, Italy
Andrzej S. Grzelakowski	Gdynia Maritime University, Poland
Mehmet Serdar Güzel	Ankara University, Turkey
Józef Hansel	AGH University of Science and Technology, Cracow, Poland
Libor Ižvolt	University of Zilina, Slovakia
Marianna Jacyna	Warsaw University of Technology, Poland
Ilona Jacyna-Gołda	Warsaw University of Technology, Poland
Nan Kang	Tokyo University of Science, Japan
Jan Kempa	Bydgoszcz University of Science and Technology, Poland
Michael Koniordos	Pireaus University of Applied Sciences, Greece
Stanisław Krawiec	Silesian University of Technology, Poland
Andrzej Lewiński	Kazimierz Pulaski University of Technology and Humanities in Radom
Bogusław Łazarz	Silesian University of Technology, Poland
Zbigniew Łukasik	Kazimierz Pulaski University of Technology and Humanities in Radom
Michal Maciejewski	Technical University Berlin, Germany
Ján Mandula	Technical University of Košice, Slovakia

Antonio Masegosa	IKERBASQUE Research Fellow at University of Deusto Bilbao, Spain
Agnieszka Merkisz-Guranowska	Poznań University of Technology, Poland
Anna Mężyk	Kazimierz Pulaski University of Technology and Humanities in Radom
Leszek Mindur	International School of Logistic and Transport in Wrocław, Poland
Maciej Mindur	Lublin University of Technology, Poland
Goran Mladenović	University of Belgrade, Serbia
Kai Nagel	Technical University Berlin, Germany
Piotr Niedzielski	University of Szczecin, Poland
Piotr Olszewski	Warsaw University of Technology, Poland
Enrique Onieva	Deusto Institute of Technology University of Deusto Bilbao, Spain
Asier Perallos	Deusto Institute of Technology University of Deusto Bilbao, Spain
Hrvoje Pilko	City of Zagreb, Department of Transport, Croatia
Antonio Pratelli	University of Pisa, Italy
Dariusz Pyza	Warsaw University of Technology, Poland
Cesar Queiroz	World Bank Consultant (Former World Bank Highways Adviser), Washington, DC, USA
Andrzej Rudnicki	Cracow University of Technology, Poland
František Schlosser	University of Žilina, Slovakia
Ewelina Sendek-Matysiak	Kielce University of Technology, Poland
Jacek Skorupski	Warsaw University of Technology, Poland
Aleksander Sładkowski	Silesian University of Technology, Poland
Wiesław Starowicz	Cracow University of Technology, Poland
Andrzej Szarata	Cracow University of Technology, Poland
Antoni Szydło	Wrocław University of Technology, Poland
Grzegorz Ślaski	Poznań University of Technology, Poland
Andrew P. Tarko	Purdue University West Lafayette, USA
Agnieszka Tubis	Wroclaw University of Technology, Poland
Frane Urem	Polytechnic of Šibenik, Croatia
Hua-lan Wang	School of Traffic and Transportation, Lanzhou Jiaotong University, Lanzhou, China
Mariusz Wasiak	Warsaw University of Technology, Poland
Adam Weintrit	Gdynia Maritime University, Poland
Andrzej Więckowski	AGH University of Science and Technology, Cracow, Poland
Adam Wolski	Polish Naval Academy, Gdynia, Poland
Olgierd Wyszomirski	University of Gdańsk, Poland
Elżbieta Załoga	University of Szczecin, Poland

Stanisława Zamkowska	Blessed Priest Władysław Findysz University of Podkarpacie in Jaslo
Jacek Żak	Poznań University of Technology
Jolanta Żak	Warsaw University of Technology, Poland
Renata Żochowska	Silesian University of Technology, Poland

Referees

Piotr Czech
Marcin Staniek
Elżbieta Macioszek
Grzegorz Wojnar
Grzegorz Sierpiński

Contents

The Digital Ecosystem in a Public Transport Company

Andrzej Kamiński(✉)

Warsaw University of Technology, Warsaw, Poland
andrzej.kaminski@pw.edu.pl

Abstract. The article presents the concept of building a digital enterprise ecosystem. The definition of the ecosystem was formulated, its attributes, and examples of application solutions were indicated. The main part of the article is a discussion of the results of a research project that was carried out in a municipal transport company. The aim of the project was to develop a digital ecosystem model. This ecosystem was built on the basis of standard IT systems supporting finance, transport and fleet management. The ecosystem project includes: building an integration platform in the operational area (automation of the process of exchanging data, messages and documents between the company's autonomous systems) and in the analytical area (data consolidation for the purpose of conducting multi-criteria economic analyses). The article presents the conclusions of the conducted scientific experiment. The experiment consisted in building a model of ecosystem in public transport company using the technology of virtual application servers and BizTalk server.

Keywords: digital ecosystem · business processes · systems integration · middleware

1 Introduction – The Concept of a Digital Ecosystem

In management sciences, an ecosystem is a structure of multilateral relationships between business partners. It is an economic community of loosely connected organizations and individuals that produce new goods and services [1, 2].

Its representative features include: focus on creating market innovations, the network structure of the organization and the changing nature of the roles performed by individual stakeholders (manufacturer, supplier, intermediary) [3]. An example of this category of structure, connecting institutional and business partners, is the innovation ecosystem. The mission of this ecosystem is the implementation of research projects and their commercialization [4]. The role of scientific institutes is to conduct research and development, construction of experimental prototypes, pilot production lines as well as verification, validation and testing of new products. The business partner participates in the financing and implementation of research works. It also commercializes, i.e. develops a prototype into a fully functional product. Then, the business partner earns revenue from the sale of products, services or licenses.

G. Sierpiński et al. (Eds.): TSTP 2023, LNNS 907, pp. 1–14, 2024.
https://doi.org/10.1007/978-3-031-53181-1_1

Information technology plays a key role in the functioning of the ecosystem. The digital ecosystem is a natural development of the concept of integrating functions, resources, processes within the organization and its external environment. It is an interdisciplinary concept, combining elements of economic sciences, management and quality sciences, as well as information technology. The main features of the digital ecosystem are: integration, symbiosis, evolution and self-organization [5]. It is a common space for cooperation, information exchange, competition and economic operations between producers and consumers. This cooperation takes place through the digitization of the physical resources of the organization and the search for new methods, technologies and communication platforms [6].

Modern ecosystems are characterized by heterogeneity. This means that various database systems (hierarchical, relational, object-oriented) are used for their technical implementation methods of data processing (transactional, analytical, heuristic) and programming techniques (structural, object-oriented, component). Currently, artificial intelligence tools (expert systems, neural networks, genetic algorithms) are also used to create this class of solutions.

The digital ecosystem can be perceived as a platform enabling the cooperation of separate, autonomous IT systems both within the enterprise and the network of economic partners [7]. The platform enables technological integration of various incompatible elements of IT infrastructure, operating systems and databases as well as software packages [8].

The digital ecosystem combines transactional, integration and analytical processes in a modern organization (Fig. 1).

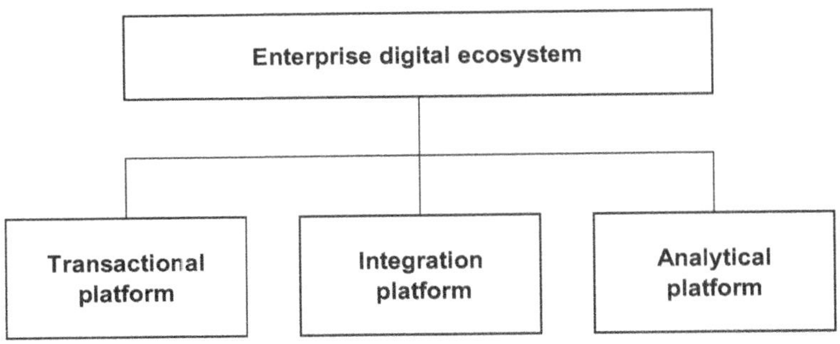

Fig. 1. Digital enterprise ecosystem (source: author)

The transactional platform enables the cooperation of business partners in order to implement a specific project. For example, an electronic auction consists in submitting successive, increasingly favorable offers by contractors. However, these tenders must comply with predefined award criteria. The essence of the auction are the evaluation criteria, according to which potential contractors are able to modify the content of submitted bids. Contractors' offers are subject to automatic evaluation based on the arithmetic formula established by the contracting authority. The criteria may be: price, order completion date or warranty period. These criteria can be expressed in terms of value or percentage.

The integration platform allows for efficient and secure exchange of information between existing IT systems. This technology is used to build bridges enabling the cooperation of transactional systems supporting financial, production or logistics management, internal and external portals and electronic data exchange systems with business partners. From an implementation technology perspective, the integration platform is embedded in the middleware layer. This allows for the construction of effective and efficient interfaces between heterogeneous IT systems of the enterprise. It is important that integration activities do not require interference with the architecture of these systems, data structures and processing technology. Middleware is software placed between a group of business applications and the operating system layer. Middleware enables the sharing of software resources between different classes of products and information technologies, and also supports work in distributed architecture [9]. Integration platforms are used in building industrial ecosystems. This concept consists in creating solutions that enable the automatic exchange of data, documents and messages between industrial automation systems, systems supporting the planning, implementation and settlement of production, as well as Manufacturing Internet of Things systems [10–12]. Middleware technology can also be effectively used as an interface between monolithic systems built in a centralized architecture and modern applications using Internet technologies and virtualization [13]. The construction of digital ecosystems is the subject of current research on industrial cyber-physical systems [14] and smart factories [15].

The analytical platform is used to integrate data from various, heterogeneous enterprise systems. The purpose of building this platform is a multi-criteria analysis of historical data to support the economic decision-making process. Source data is consolidated in a data warehouse structure. A data warehouse is a repository designed to collect, standardize and aggregate copies of transactional data. This data is recorded, processed and stored in various, autonomous business systems and applications. OLAP (Online Analytical Proceedings) technology is used to optimize the processing of large sets of analytical data. In contrast to the relational model paradigm, which is normalization, the source data and aggregates are stored in a multidimensional structure. The cost of this solution is data redundancy. However, the use of OLAP technology enables the creation of cross-sectional statements and reports in real time.

Managing a modern transport company requires the implementation, maintenance and development of various, autonomous IT systems supporting passenger transport planning, vehicle fleet management, or multi-criteria analysis of bus, staff and infrastructure maintenance costs. The goal of building a digital ecosystem in public transport company is the cooperation of many heterogeneous applications at the transactional, analytical and integration level. This solution will be flexible, cost-effective and able to quickly adapt to organizational changes. The ecosystem is an alternative to multifunctional, monolithic, customizable integrated systems.

2 Model of the Ecosystem in Public Transport Company

The research was carried out in the municipal transport company (MTC). The subject of the research was the analysis of the flow of data, documents and decisions between individual organizational departments of the company. The functionality, technical architecture and the possibility of servicing and developing existing systems were also assessed.

On the basis of the conducted analytical work, functional requirements for the future, comprehensive IT ecosystem were defined.

MTC used several applications supporting the basic business areas: finance and accounting, logistics, human resources management, renovation and repair management, planning and settlement of passenger transport.

Theoretically, the simplest solution would be to implement a centralized, integrated IT system supporting the above-mentioned functional areas. This system would replace many technologically outdated applications. The potential economic benefits resulting from the implementation of the integrated system include: elimination of data redundancy, reduction of operating costs related to multiple inputs of the same data to various applications and the ability to create cross-sectional statements and reports for the needs of the management board.

However, based on the analysis of the market of systems supporting the management of a municipal transport company, it turned out that the national IT solutions do not meet the functional requirements of the examined company. It was also found that the adaptation of foreign solutions is technologically risky and economically unprofitable. Adaptation is defined as the process of reconfiguration of standard software to the specificity of national laws and regulations. In practice, this means the need to interfere with the system kernel and the need to analyze and reconstruct thousands of lines of application code.

An alternative solution is to build an MTC ecosystem based on three standard systems supporting the financial, transport and transportation areas. On the other hand, Business Intelligence class software will be used for the purposes of strategic management (Fig. 2).

In detail, the MTC ecosystem will be composed of four independent systems::

1. ERP system supporting logistic, financial and accounting and personnel processes;
2. Transport management support system - registration, planning and settlement of passenger transport, as well as planning tasks and working time of drivers;
3. Fleet management support system - register of vehicles and their equipment, control of subassemblies, planning inspections, repairs and maintenance;
4. Business Intelligence - multi-criteria analysis of the operating costs in public transport company.

The foundation of the MTC ecosystem will be the ERP software package. This system will support typical areas of activity related to: financial management, human resources management, and fixed assets management. ERP will cooperate with specialist systems, i.e. a system supporting the management of passenger transport and a system supporting the management of a fleet of vehicles. Each of these systems will be a specialist, multi-module software package embedded in a separate database. These systems come from different manufacturers. They were built using mutually incompatible development tools and platforms. In Fig. 2, the autonomy of specialized systems is symbolized by separate, unconnected cubes. As part of these systems, financial, logistic and transport operations will be carried out (transaction platform). The benefits associated with the use of standard software packages include: functional advancement, technical quality and a guarantee of continuous product development. For the most part, standard systems are devoid of annoying bugs. For example, typical software errors include the lack of full control of the correctness of the entered data. The reason for this type of errors is usually

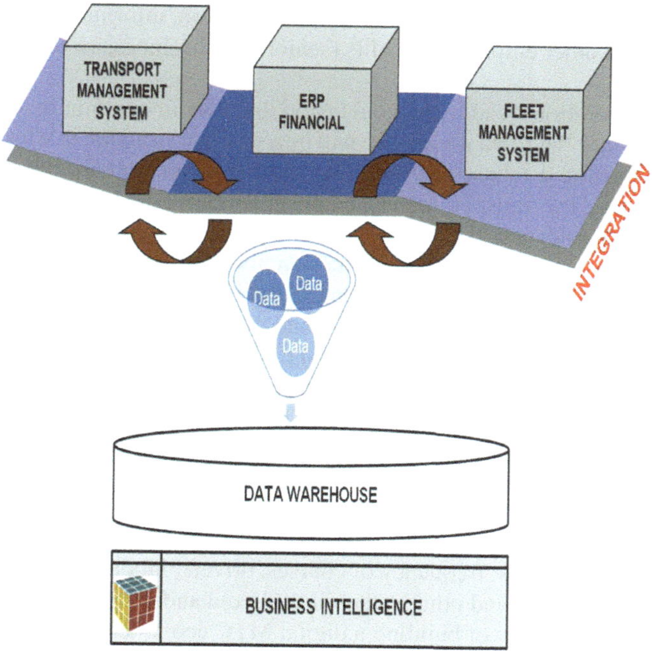

Fig. 2. MTC ecosystem model (source: author)

the inability to fully check all possible user actions. These errors are characteristic of author's software, which is created at the request of a given organization. It should be noted that the implementation of the listed standard packages does not yet lead to the creation of an enterprise ecosystem. The ecosystem is a synergetic solution in which the heterogeneity of individual applications is hidden. Therefore, it is necessary to develop integration solutions enabling the cooperation of the above-mentioned systems both at the operational and analytical level.

Data and electronic documents will be exchanged via middleware. In principle, the cooperation of IT systems at the transaction level will enable the automation of business processes in a public transport company. In Fig. 2, the middleware layer is presented as the platform that is the foundation of the MTC ecosystem. The company's IT systems are located on this foundation. Semi-circular arrows symbolize data flow between applications.

Integration at the operational level will enable process automation and eliminate data redundancy. However, building an integration platform will not solve the problem of data consolidation for analytical processing. For example, the history of service and repair orders is available in the system supporting fleet management. Based on the analysis of these data, the labor consumption of individual service operations can be determined. The analysis of the order history can also be useful to determine the hourly cost of technical employees dealing with fleet maintenance. However, these data do not show the full spectrum of labor costs for service personnel. There is no information on costs related to planned layoffs, holidays or training. In addition, information on costs derived from salaries and other employee benefits (seniority allowances, awards, lump sums) is also unavailable.

These considerations lead to the conclusion that data integration for analytical processing is necessary. The data warehouse will be a separate structure in which operational data from transaction databases will be collected. These are data from the financial and accounting system, the transport management support system and the fleet management support system. This data will be extracted from the source systems, integrated and then entered into the structures of the central data warehouse. For the purpose of multi-criteria data analysis, the Business Intelligence technology will be used in MTC. This technology is used to process large volumes of data and their presentation and publication. For the purposes of data visualization, managerial dashboards and multi-criteria summaries and reports are used. For example, dashboards provide: tabular data, charts, maps, infographics, scorecards and predefined reports. These tools enable the construction and monitoring of KPIs (Key Performance Indicators) in the enterprise.

The final effect of the implementation of the MTC ecosystem construction project will be the development of an analytical model supporting the control of vehicle-kilometre production costs. The model will take into account the parameters concerning the type of transport routes, frequency of courses, drivers' labor costs, costs of purchase and operation of the fleet and other categories of fixed and variable costs.

To sum up, the process of building a digital MTC ecosystem can be conventionally divided into three stages:

I. Implementation of standard software in the area of finance, transport and fleet.
II. Building a platform for integrating transactional systems.
III. Building an economic analytics platform.

A detailed discussion of the listed stages of the project is beyond the scope of this article. This is due to the complexity of the issue and the need to use various methods, tools and information technologies. It should be noted that the implementation of systems in the area of finance, transport and fleet should be carried out in accordance with the supplier's methodology. Work related to the installation, parameterization and configuration of the standard software should be regarded as routine work. Typically, these works are carried out in accordance with a strictly defined implementation procedure.

The implementation of autonomous systems in the area of finance, transport and fleet does not yet create a digital enterprise ecosystem. In the author's opinion, the integration platform is a key element enabling the automatic exchange of data and electronic documents between MTC transactional systems. Therefore, in the further part of the article, the author's methodology for building an integration platform will be presented, and adequate application solutions will be indicated.

3 Description of the Experiment – Building and Testing the Integration Platform

Building an IT systems integration platform requires developing an appropriate project management methodology and conducting a scientific experiment.

The experiment consisted in developing a prototype of the MTC systems integration platform. These works were carried out in laboratory conditions. Tests of data and document flow between autonomous MTC systems were carried out using virtual application servers.

Virtualization is the process of creating a "virtual" version of a physical computer with a certain number of processors, a volume of main memory and permanent data storage. The process of migrating a virtual machine to a physically separate computer means distributing a few selected files. This saves time and costs, as the process does not require time-consuming installation and configuration of the operating system and other software packages. In research and development projects, virtualization technology enables simple exchange of newly developed software components between individual team members. The benefit of using this solution is the elimination of technical work related to the migration of the test environment and system settings.

The experimental solution was developed by the author using the ARIS-Toolset software package and the MS Biz-Talk integration server.

Business process models were developed using the tools available in the ARIS-Toolset package. The basic features of this package include: graphical modeling of functions, data, processes (ARIS Modeller), analysis of process time and costs (ARIS Analyzer), as well as management of the central repository of models (ARIS Navigator). ARIS-Toolset enables simulation research. The purpose of these studies is to identify dysfunctions in the existing flow of data, documents and decisions, taking into account time and costs. Based on the results of simulation studies, it is possible to develop and implement appropriate improvements.

The integration platform prototype was implemented using the MS Biz-Talk server. This tool enables the design and automation of processes related to the exchange of electronic documents. In the BizTalk Server environment, it is possible to define input and output data structures and data transformation schemes. The integration project includes the creation of organization models and IT systems models as well as the configuration of channels and ports for the transmission of electronic documents. The basic medium for data exchange are files written in XML (eXtensible Markup Language). BizTalk has interfaces (accelerators and adapters) to standard business applications (e.g. SAP ERP) and database management systems.

The experiment consisted of two stages, i.e. development of a new configuration of business processes and implementation of the integration platform.

The first stage included the development of a MTC process model. As part of the analytical work, data, documents and decisions were identified. In particular, the following were analyzed and assessed: the organizational structure of the company, workflow, and the architecture and functionality of the existing IT systems. Then, a model of processes in the area of transport, fleet and finance was developed using the ARIS-Toolset package.

The second stage involved the physical launch of the flow of electronic documents in the integration platform layer. In technical terms, any of the three. MTC's key IT systems (finance, transport, fleet) were installed and configured on a separate virtual machine. The fourth virtual server was dedicated to the implementation of the integration platform. The model of business processes was mapped in the integration platform layer. On this basis, the flow of electronic documents was configured and the appropriate integration infrastructure was prepared (channels, ports, adapters).

The aim of the experiment was:

1. Analysis of the technical conditions for building a digital ecosystem in a public transport company.
2. Studying the level of automation of business processes. Analysis of the process of exchanging electronic documents between autonomous IT systems. Evaluation of integration technologies, i.e. transactionality and security of processing operations.
3. Assessment of solution flexibility. Analysis of the ease of introducing modifications and changes to the structure of electronic documents and process models. It was assumed that these changes should not require time-consuming and cost-intensive programming work.

The scientific result of these works is the development of a methodology for building a digital ecosystem of MTC in the area of transaction systems integration.

4 Methodology of Building a Digital Ecosystem – An Integration Platform

Building a digital ecosystem in a public transport company requires analytical, conceptual and technical work (Fig. 3).

Stage 1. Conceptual Design of the Integration Solution

The implementation of the platform should be preceded by the development of a new configuration of business processes. For example, the transport process includes: planning drivers' working time (taking into account holidays, holidays, sick leave, etc.), implementation of commissioned tasks, and then their monthly settlement. The implementation of this process requires the development of integration solutions. It is necessary to automate the flow of data, documents and decisions between the ERP system (finance and accounting module, HR module) and the system supporting transport management. A separate problem is the process of integration between the transport management support system (planning the daily demand for rolling stock) and the fleet management system (planning inspections, repairs and other service activities).

Decisions will include the following issues:

1. In which system will the central database of drivers' personal data be kept?
2. Will the transport system record only drivers' working time, or will their monthly wages also be calculated?
3. What information about the structure and technical and operational characteristics of vehicles should be replicated between the transport system and the fleet management system?

Formulating answers to these questions requires the development of business process models at MTC. This work includes functional decomposition, i.e. defining the main processes, followed by the specification of sub-processes, procedures, objects and actors. The result of this work is the development of a static process model. This model illustrates the flow of information and logical connections between individual objects of the organization. The static model can be effectively used to carry out simulation studies. However, this model does not allow for the construction of physical connections between the integrated systems. The implementation of an integration solution will require the development of a dynamic model of processes, which will be discussed later in the article.

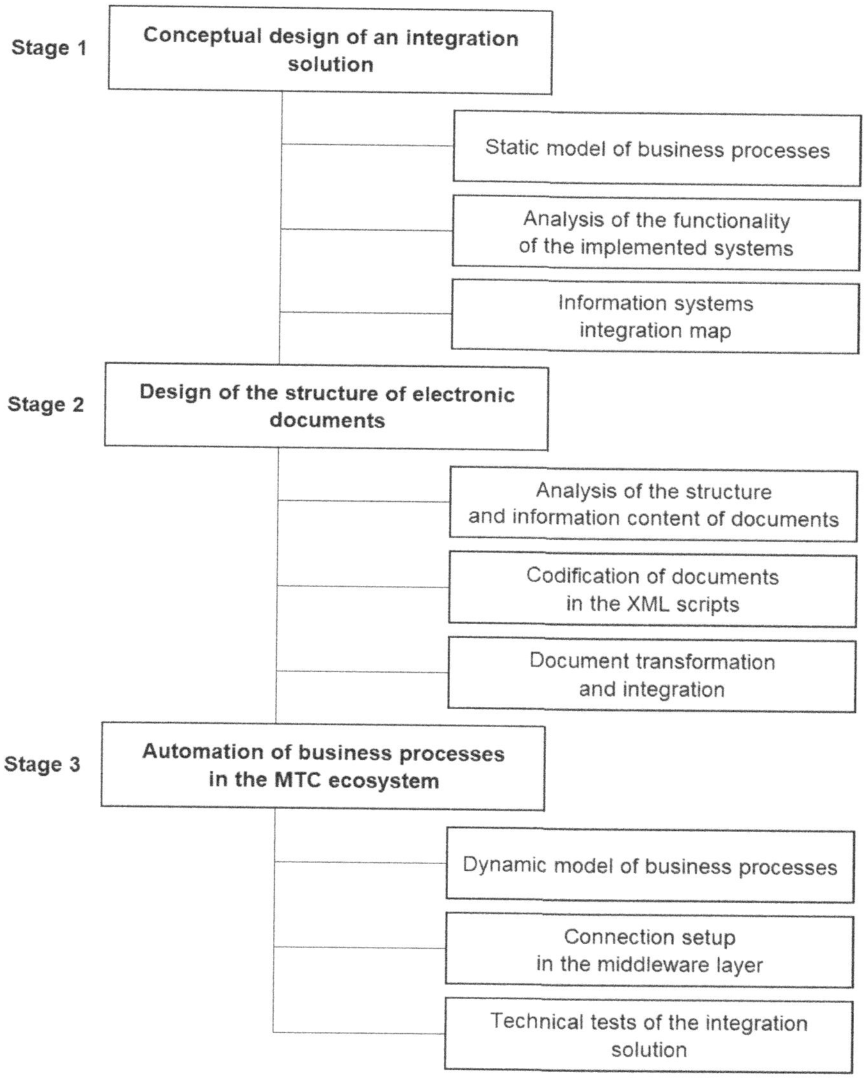

Fig. 3. Model of IT systems integration in MTC (source: author)

As part of conceptual work, it is also advisable to analyze the functionality of previously implemented IT systems at MTC. This analysis should include the specification of documents to be exchanged between these systems and the definition of objects (departments of the organization) and actors (i.e. persons responsible for document registration, acceptance and decision making).

The conceptual design should include a specification of reference connections between IT systems supporting MTC management. Ultimately, a map of enterprise systems integration will be created. Existing and newly designed IT systems and reference connections between them will be marked on this map. The purpose of this work is to divide tasks and processing functions between standard ERP software modules (finance and accounting, human resources and payroll, fixed assets, etc.) and specialist systems supporting transport and fleet management.

In order to enable the transmission of documents in electronic form, it is necessary to use a specific method of recording standardization. It was assumed that the basic technique for data exchange will be an electronic document. When constructing electronic documents, the XML language is used. This language is used to record information content, document structure and metadata. XML is a standard for codifying data (documents) distributed and published on the Internet.

The conceptual design should also take into account the analysis of data sets subject to exchange between integrated IT systems. The structure of documents should be subjected to a detailed analysis, data formats should be agreed and the method of transformation of individual data fields should be determined. For example, the employee file is created in the HR (ERP) system, as well as in the system supporting transport management. It is therefore necessary to determine which of them will act as the "master system" and which will be the "partner system". The master system will record personal data and any employee events. Data and documents related to the personnel service of drivers will be manually entered into this system. Entering a specific set of data into the master system will be associated with the transfer of the appropriate document to the partner system. Registration of the driver's leave application in the HR (ERP) system will result in issuing the appropriate document, which will then be automatically loaded and processed in the transport system.

To sum up, work on the conceptual design of the digital ecosystem should be initiated by a process analysis of the company. Development of the process design, its graphical representation, as well as modeling and optimization of the flow of information and decision-making resources is a task for business analysts. The result of the analytical work will be the development of a document flow model and the determination of the rules for their recording and processing in MTC IT systems. The results of this work will be the basis for developing a map of information systems integration. This map will illustrate the static flow of data, documents and decisions.

Stage 2. Designing the Structure of Electronic Documents

The medium for data exchange between the integrated systems will be an electronic document. In the process of creating the structure and information content of these documents, the following can be distinguished: logical design and physical design.

The logical design is the result of analytical work, which includes: verification of the information content of individual documents, specification of attributes and data types, as well as a description of validation rules.

The physical design includes the codification of documents in the form of XSD schemas (XML Schema Definition) in accordance with the W3C (World Wide Web Consortium) specification. In other words, electronic documents will be saved in the form of a graphical, hierarchical tree of nodes and application code, i.e. schemas and XSD scripts.

In the BizTalk development and design environment, a new document definition can be created from scratch or imported from an XML file. The document definitions will be saved in the BizTalk Server repository. Vision tools are used to design the structure of new documents. It is also possible to edit and introduce modifications, corrections and changes to existing documents. In special cases, changes in individual XML scripts are introduced manually (modification of the code).

The next step is the transformation and integration of documents from various sources. Let's consider this issue on the example of a driver's leave application. Let's assume that this application is registered in the HR system (master system). On this basis, an electronic document should be automatically created, which - using the integration platform - will be entered into the transport system. Integration activities will include mapping the fields of the leave application document, i.e. building connections between individual fields in the source file and in the output file. BizTalk Server also has special mapping functions that can be used to define complex dependencies between fields. For example, using the conversion function, it is possible to automatically extract an employee's date of birth based on their PESEL number.

Summing up, the result of this stage of work will be the creation of a central repository of electronic documents. This repository will be placed in the integration platform layer. Transformations will be developed to enable the transfer of documents between heterogeneous MTC IT systems.

Stage 3. Automation of Business Processes in the MTC Ecosystem

The dynamic model of the business process will be implemented in the integration platform layer. Process implementation includes configuration of channels, communication ports and adapters. The data exchange medium will be electronic documents. The internal logic of the process will be saved in the integration platform layer.

BEPL (Business Process Execution Language) will be used to translate the process model to the integration platform environment. BEPL is a declarative XML markup language for defining processes based on web services. This language enables the construction of complex processes using classical algorithms and software structures. Using the algorithm, subsequent sequences of actions in the process can be determined. It is also possible to define conditional clauses. This means that the results of the work of the "X" process can be arguments of the variables that initialize the execution of the "Y" process. BEPL also offers traditional programming constructs such as: variable declarations, loops, event and exception handling. BPEL is a network service orchestration language. In orchestration, the service communicates only with the orchestration manager, in which logical connections between the integrated IT systems have been defined. In the MTC ecosystem project, the BizTalk integration server is the orchestration manager.

Saving the process model in the BPEL language enables the flow of electronic documents between MTC's autonomous IT systems. Let's assume that a plan for periodic inspections, repairs and overhauls of buses is created in the fleet management system. As a result, it is necessary to develop a demand for parts, subassemblies and other components necessary to carry out these works. The relevant documents are created in the fleet management system. On the other hand, the process covering: registering suppliers, placing orders, registering contracts, handling settlements and warehouse management will be carried out in the ERP system. Electronic documents will be automatically transferred between the fleet management support system and the ERP system. Note that this is a two-way process. In order to carry out current work orders, purchased parts, subassemblies and consumables will be successively taken from the warehouse. The warehouse issue document will be registered in the ERP system. This action is necessary to properly keep records of costs and revenues in MTC. According to the previously designed process model, this document will be automatically emitted from the ERP system and then uploaded to the fleet management system. In the latter, a virtual warehouse will be created, which will be used to issue parts for individual work orders. This warehouse is necessary to correctly associate parts and materials with work orders issued for the repair of individual buses. It is also advisable to record the dismantled subassemblies. These elements (after repair or regeneration) can be used to modernize another bus. Full records of service and repair activities will be kept in the fleet management system.

Based on the above description of the process, one can get the impression that there is still duplication of warehouse operations. It should be noted that in order for the above-mentioned systems to work properly, appropriate storage facilities must be established in each of them. However, these objects have a separate purpose. In the ERP system, the warehouse is an essential element of the logistics process related to handling orders and financial settlements. On the other hand, in the fleet management system, the warehouse is a component of the process related to the ongoing operation, service and repair of buses.

Based on the conducted experiment, it was found that the existence of duplicate storage facilities is necessary to maintain the correct operation logic of these systems. An important question therefore arises: What is the real benefit of implementing an integration platform?

This advantage is the principle of once data and document entry within the entire MTC ecosystem. For example, a requisition document for specific materials and parts is created in the fleet management system. Creating a new data record initiates the action of issuing an electronic document. This document is processed in the integration platform layer. Then, a transaction is automatically triggered in the associated ERP system. This document is entered into the system database and is subject to further processing in the process of servicing and settling purchases. Thus, information redundancy in the ecosystem has been eliminated. To put it simply, each document is entered only once into the appropriate IT system. The integration platform controls the automatic flow of electronic documents in the MTC ecosystem.

5 Summary

The problem of systems integration becomes crucial in the case of heterogeneous solutions. The MTC ecosystem will not be a homogeneous system embedded in a central database. Therefore, the design of integration solutions is of great importance.

Based on the conducted experiment, it was found that the implementation of a new configuration of business processes will eliminate data redundancy. This means that documents will be entered into the ecosystem once, in accordance with the developed process model. In turn, the integration platform will enable the automatic exchange of electronic documents and the transfer of decisions. The heterogeneity of standard systems will be hidden.

However, in the author's opinion, the final verification of the developed and implemented new configuration of business processes will take place only at the stage of production launch of the MTC ecosystem. Therefore, the flexibility of the integration solution, i.e. the ability to redefine the implemented business processes, is extremely important.

Building the MTC ecosystem included the integration of internal processes. However, integration activities may also concern building links between IT systems of different enterprises. The MTC ecosystem should be open and cooperate with the IT systems of contractors, cooperators and business partners.

References

1. Adner, R.: Ecosystem as structure: an actionable construct for strategy. J. Manag. **43**(1), 40 (2017)
2. Senyo, P.K., Liu, K., Effah, J.: Digital business ecosystem: literature review and a framework for future research. Int. J. Inf. Manag. **47**, 3 (2019)
3. Pidun, U., Reeves, M., Schüssler, M.: Do you need a business ecosystem. BCG Henderson Institute, pp. 2–3 (2019)
4. Oh, D.D., Phillips, F., Park, S., Lee, E.: Innovation ecosystems: a critical examination. Technovation **54**, 1–2 (2016)
5. Senyo, P.K., Liu, K., Effah, J.: Towards a methodology for modelling interdependencies between partners in digital business ecosystems. In: IEEE International Conference on Logistics, Informatics and Service Sciences, (LISS). IEEE (2017)
6. Subramaniam, M., Iyer, B., Venkatraman, V.: Competing in digital ecosystems. Bus. Horiz. **62**(1), 84–86 (2019)
7. Barykin, S.Y., Kapustina, I.V., Kirillova, T.V., Yadykin, V.K., Konnikov, Y.A.: Economics of digital ecosystems. J. Open Innov. Technol. Market Complex. **6**(4), 5–8 (2020)
8. Behera, P., Dash, C.: Digital ecosystems: challenges and prospects. Int. J. Res. Anal. Rev. **6**(1), 179–181 (2019)
9. Yongguo, J., Qiang, L., Changshuai, Q., Jian, S., Qianqian, L.: Message-oriented middleware: a review. In: 2019 5th International Conference on Big Data Computing and Communications (BIGCOM), pp. 88–90. IEEE (2019)
10. Alfalouji, Q., et al.: IoT middleware platforms for smart energy systems: an empirical expert survey. Buildings **12**, 526 (2022)
11. Vikash, Mishra, L., Varma, S.: Middleware technologies for smart wireless sensor networks towards internet of things: a comparative review. Wirel. Pers. Commun. **116**, 1539–1574 (2021)

12. Benayache, A., Bilami, A., Barkat, S., Lorenz, P., Taleb, H.: MsM: a microservice middleware for smart WSN-based IoT application. J. Netw. Comput. Appl. **144**, 138–154 (2019)
13. Caballero, V., Valbuena, S., Vernet, D., Zaballos, A.: Ontology-defined middleware for internet of things architectures. Sensors **19**(5), 1163 (2019)
14. Stock, D., Schel, D., Bauernhansl, T.: Middleware-based cyber-physical production system modeling for operators. Procedia Manufact. **42**, 111–118 (2020)
15. Zhang, L., Yuan, H., Chang, S.H., Lam, A.: Research on the overall architecture of internet of things middleware for intelligent industrial parks. Int. J. Adv. Manuf. Technol. **107**, 1081–1089 (2020)

Contracts of Transport Services Provision as a Factor of Increasing the Effectiveness of the Urban Public Transport

Grzegorz Dydkowski(✉), Wiktoria Gałecka, and Anna Urbanek

Department of Transport, University of Economics in Katowice, Katowice, Poland
{grzegorz.dydkowski,wktoria.galecka,anna.urbanek}@ue.katowice.pl

Abstract. Public financing of urban collective transport enables the provision of urban collective transport services where the revenues from the sale of services do not cover the costs, but it is justified by social reasons and the implementation of a pricing policy that increases the availability and improves the competitiveness of collective transport in relation to travel by passenger cars. It also ensures the stability of service provision and favorable conditions for the purchase of modern low-emission rolling stock and other investments.

The paper, based on the experience in the operation of the urban collective transport in the central part of the Silesian Voivodeship, presents the direction of their evolution, so as to reduce the operators burden of the risk of events, which they cannot affect, but at the same time, only to a limited extent, to introduce risks and thereby an element of remuneration due to the services provision, depending on the transport volume. It is expected, that, together with relevant procedures for operators to submit conclusions and suggestions of changes in the transport offer, it will result in the increased productivity and rationality of the services provision by the urban public transport in cities.

Keywords: urban transport · contracts · risk allocation

1 Introduction

Pursuing the effective spending of public funds assigned to the urban public transport, and obtaining benefits from the whole city and its residents point of view, the market orientation of entities in the urban transport is necessary, their activation aimed at the provision of high quality services, and through that at the increased volume of transport and the public transport's share in the transport service of cities. This may be achieved in various ways, e.g. via proper division of competences and tasks in the organisation and provision of public transport services, and implementation of mechanisms, which motivate to undertake projects directed at the increase in the transport volume. The passivity in this field, both of organisers and carriers, will result in a declining transport volume and thereby in the need for increased public financing and effects related to excessive development of individual motorisation.

G. Sierpiński et al. (Eds.): TSTP 2023, LNNS 907, pp. 15–33, 2024.
https://doi.org/10.1007/978-3-031-53181-1_2

In the last decades of the 20th century and in the first decade of the 21st century the procedures of competitive conclusion of contracts with operators were the factor intended to increase the economical spending of public funds in the urban public transport, however, this tool, in particular in the European Union states, is less and less often used. The taking effect of Regulation (EC) no 1370/2007 of the European Parliament and of the Council [1] in 2009 enabled direct conclusion of contracts with entities under public control, albeit the decision to apply a competitive procedure or direct conclusion of contracts, and the contract contents, including service standards, was left to competent authorities of the EU Member States [2]. However, not denying the adopted solution, in which the direct contracts conclusion is possible and the fact that in the urban public transport the public intervention is necessary, because in many cases the services cannot be provided based on commercial rules, it is necessary to state, that it facilitates the conclusion of contracts without competitive procedures and market verification of the costs level, which results next in maintaining the public ownership on the level of operators carrying out transport. Competitive procedures, preceding the contract conclusion, are considered one of factors, which force the cost controlling actions [3–6]. The solutions, in which the public ownership and public financing dominate in the provision of collective transport services, are considered less economical than those with the participation of the private sector, and a part of public financing covers the increased costs existing in public systems [7]. In the situation of public financing, of public ownership of entities carrying out transport, and also of a direct selection, without the application of competitive procedures for contractors, this means the necessity of developing new solutions, aimed at the motivation for rational and effective provision of services. Otherwise, in a longer perspective it is necessary to consider increasingly great, over time exceeding city possibilities, demand for public funds feeding the public transport. However, the impact of the mentioned regulation on the economic and financial aspects of the sector, as well as the implementation of transport policy goals, is more complex [8].

2 Objective and Research Methods, Research Problems

One mechanism that can improve the effectiveness of urban collective transport is through contract provisions, particularly by specifying the scope of duties, risks, and the remuneration system. However, there may be different limitations and roles involved [9]. The contracts, concluded by the public authorities responsible for the ensuring of urban public transport with entities providing such services, may be systematised depending on the degree of public authority interference, from the services provision based on market rules, up to highly regulated systems. In big cities in the European Union states the management of public transport systems by the organisers, and a model of contract with operators in so-called gross costs formula, are quite common. Despite numerous advantages of this solution, its drawback consists in the fact that the operators' remuneration does not directly depend on the volume of passenger transport, which weakens the system effectiveness. The operator carries out the operational work in accordance with timetables prepared by the organiser, frequently they also expect timetables convenient from the point of view of transport organisation itself and the costs minimisation, and not such, which are primarily directed towards the maximisation of the transport volume. The net costs contracts, i.e. such in which the carrier is burdened with the risk of revenue on sales of services, are generally in smaller cities, for lines connecting cities with suburban areas, and in suburban areas.

The objective is to present two types of contracts used in the urban public transport, that is the net costs contracts and gross costs contracts, the systematics of tasks and risks allocated to entities, the contracts assessment from the point of view of directing the service providers at the increase in this sector effectiveness, services quality, number of carried passengers, and thereby at increasing the utilisation of the transport offer. The number of carried passengers should increase primarily through the improvement in quality and better adaptation of the offer to the needs. Based on that the directions of contents modifications will be indicated for the contracts of transport services provision concluded by organisers with operators.

The research on the entities behaviour utilised methods of observations and discussions with the management and employees of entities in the urban public transport, and with representatives of cities in the central part of the Silesian Voivodeship, of documents analysis, in particular of contracts, and also the analysis of the subject literature. A principle was adopted during the synthesis, that solutions should be shaped so, that the tasks and risks were assigned in particular to the party, which can affect them, and thereby it can better manage them.

Qualitative methods were used during the research, despite the fact that their usefulness may not be valued in certain research circles. It is necessary to point out that the qualitative studies are devoid of objective valuation of the existing reality, although in many cases, in methods based on systems of measures and on the collected data, it is also possible to have comments on the quality of information, measurement methods, and the scope of events, which a given measure represents, reflects, or evaluates. Approving the high importance of quantitative methods and the current domination of quantitative studies in economic sciences, the attention should be drawn to the fact that by quantitative studies the answer is obtained primarily to questions related to the number, frequency of solutions, dynamics of events over time, or conclusions are drawn based on the studied sample of the whole population. The quantitative studies pursue to determine the quantitative relationships between the studied phenomena, paying attention primarily to the correctness of determined variables and the logic of measurements [10]. As a result, this allows to determine facts, to quantify the existing situation, and to check various relationships, however, their application is limited at the development of theories and formulation of suggestions [11]. But the qualitative studies are equally important, they less focus on the frequency of studied solutions or phenomena, and more on the description, presentation, and development of theories and related best practices. In this case answers are obtained to basic questions - why and how a given phenomenon occurs, and the understanding and determination of such actions and behaviour motives is pursued [12]. The qualitative studies allow to get deep into the analysed processes [13–16]. The qualitative studies enable to recognise many details in attitudes and behaviour, which are difficult to identify using other methods, allow to approach holistically complicated and dynamic social phenomena, and also allow to study and conclude based on various cases, not necessarily typical, but just separated. Because in quantitative studies only typical things are considered, frequently the elimination of extreme, different data is a normal practice, not to deform the calculated statistical measures [13, 17]. The developed theory is to certain extent general, and with reservations related to its practical application. The basic role of qualitative studies consists in the development of a theory, utilising at that the experience of researchers participating in the project. It is considered that qualitative methods have a significant potential, because, using the data and examples of social processes, they can provide a lot of information and clarifications for the formulated theories, and also can enable to generalise theories to these elements of phenomena, which are not subject to quantification [18]. So it is possible to assume theoretical testing, with the use of case studies. It is worth drawing attention to the fact, that the qualitative studies are not easier to conduct, than the quantitative ones, and in the past they fulfilled and continue to fulfil an important role in cognitive processes.

3 Risk Division Between the Public Transport Organiser and Operator in Contracts of Net Costs and Gross Costs

The division of risk between parties to the contract should be a derivative of the division of tasks and duties; the entity responsible for a given task should also incur the risk related to its non-performance. Hence the risk should be assigned to the entity, which can manage it, identifies and can affect the risk factors, can and is interested in preparing to unfavourable circumstances, and also relatively quickly obtains the information about their occurrence. Table 1 presents the existing division of risks in contracts between the organiser and operators providing the transport services.

Table 1. Risks allocation between the organiser and operator

Type of contract concluded between organiser and operator	Entity incurring specific risks	Risk description
Gross costs contracts	Organiser's risks	- risk of the planned size of demand, and ultimately of the income on sales of services, - risks related to possible irregularities in the provisions of or in the resolving of tariff regulations, - risks related to not ensured accessibility to the sales network, inspections, and effective recovery of additional charges, - risk of missing public funds at the amount previously planned to maintain the transport offer for the period of resolution validity, - risk related to introduced changes in the transport offer and in tariff systems, - risk of imprecise or too general provisions of the contract, so that it would reflect the objectives of the transport policy, its flexibility, and possibility of introducing changes and innovations

(*continued*)

Table 1. (*continued*)

Type of contract concluded between organiser and operator	Entity incurring specific risks	Risk description
	Operator's risks	- risk of exceeding own costs and, as a result, impossibility of covering them in full from the payment for the provision of services obtained from the organiser, - risk of adopting adjustment formulae, which will not reflect the actual increase in costs, which for long-term contracts can result in financial deficits, - risk of impossibility of ensuring resources necessary to provide services, in particular vehicles and human resources (in particular shortages in the recruitment of vehicle drivers), - risk of decrease in the orders for transport and related failure to utilise the transport potential, or demand for transport by vehicles other, than those possessed by the operator
	Common risks	- risk of contract conclusion in accordance with the binding legal regulations, - risk of contract provisions ambiguity and related disputes during its execution, - risk of missing provisions in the contract in the field of part of tasks and duties and related remuneration, - risk of difficulties in the vehicles traffic and related quality deterioration of transport services, - risk of force majeure occurrence and related difficulties in the services provision, of increased costs, or just necessity for changes in the rules of their provision

(*continued*)

Table 1. (*continued*)

Type of contract concluded between organiser and operator	Entity incurring specific risks	Risk description
Net costs contract	Organiser's risks	- risk of ensuring an appropriate amount of public funds for subsidies to services provided by carriers, - risk of services provision by the carrier in a way, which takes into account objectives of the transport policy, - risk of transport offer reduction on lines and in areas with low demand, and the implementation of welfare and social functions through prices, - risk related to the limitation of common tickets offer
	Operator's/ carrier's risks	- risk of demand for services and, as a result, covering own costs with income on tickets sales and financing from public funds, - risks related to ensuring the accessibility to the sales network, inspections, and the effective recovery of additional charges, - risk related to changes introduced in the transport offer and in tariff systems, - risk related to the provision of services and competition of other carriers in the entire area, its part, or on selected routes or connections, - risk of difficulties in the vehicles traffic and related quality deterioration of transport services

(*continued*)

Table 1. (*continued*)

Type of contract concluded between organiser and operator	Entity incurring specific risks	Risk description
	Common risks	- risks related to the correctness of tariff regulations, - risk of contract conclusion in accordance with the binding legal regulations, - risk of contract provisions ambiguity and related disputes during its execution, - risk of missing provisions in the contract related to a part of tasks and duties and the related remuneration, - risk of force majeure occurrence and related difficulties in the services provision, of increased costs, or just necessity for changes in the rules of their provision

Source: own study

However, the risk division between the parties to the contract not always takes into account the rule pursuing the reduction of total risk costs, and after all that should be in the interest of the party ordering and financing a specific service or project. Because frequently the employer, in this case the transport organiser, transfers on the contractor also these risks, which the contractor cannot influence. The basis of such behaviour consists in the intention to reduce the risk carried out by the employer, hence the pursuit to get rid of the risk, caused by the unwillingness to manage the risk and thereby, if it materialises, the events in their nature difficult and resulting in higher or lower losses. The contractor, in the case of contractual provisions, from which risks result, which it cannot affect and which do not depend on him, most frequently evaluates them and includes in the cost related to the contract implementation, which ultimately raises the total costs, and hence the employer's expenditures, which are higher than if the risk is placed in a way, which enables its management.

4 Discussion and Results

Appropriate risk allocation, i.e. identification and clear and realistic (optimal, appropriate) allocation to each partner of the responsibility for negative effects resulting from the occurrence of a specific risk factor, is crucial for minimisation of the risk and negative effects of its materialisation. The assumption, that the effects of a given negative event occurrence should burden the party, which can affect the probability of its occurrence, which can monitor the risk, and also relatively early acquire the information about the risk transformation into the occurrence, should be a proper principle. As a result, it can better protect itself against the undesired occurrence and its negative effects, and take actions faster, thereby reducing losses. This entity can reduce the risk and at the same time it is interested in that, because primarily it will be burdened by negative risk effects.

It is not proper to be guided by a rule, in which the risks are transferred to the contractor to the maximum extent and even for any price. Because, from the risk management point of view, both minimisation of the probability of undesired situations occurrence, and minimisation of negative effects in the case of materialised risk are crucial, and just that is ensured by appropriate risk allocation. The appropriate risk allocation provides benefits from the entire project point of view, it allows to reduce the amount of losses due to undesired situations, the loss calculated for contract parties altogether. The appropriate risk allocation means also better cooperation during the contract implementation, and a lower probability of disputes or of the occurrence of a situation, in which the contract will be terminated before the end of period, for which it has been concluded, and before the performance of the work, which was its subject. Disputes also engage resources of entities and also result the reduced involvement in the very project, the contract fulfilment.

Because of, despite certain limitations, discretion in the development of contractual provisions, and hence various provisions in the contracts, the risks allocation in contracts in many cases, considering the reduction of negative risk effects, is far from optimal. Because there exist situations, in which irrespective of the fact, whether a given party to the contract can affect a specific occurrence or not, it has the risk allocated in the contract, in relation to which it has limited possibilities of managing it, and this ultimately means, that it is difficult to reduce the financial results, when the risk converts into occurrence. The lack of proper risk allocation to parties to the contract can result from various reasons. In part of cases the employer, the public transport organiser, tries to minimise the risk incurred by itself and transfers it to the contractor, the operator providing the transport or another entity, who, during the estimation, in certain cases includes it in the bid price. The transfer of as high as possible risk to the contractor occurs in particular in the case of public entities, where the employer prepares the participation conditions, bids evaluation criteria, and a draft contract or general conditions of the contract. Potential bidders can obviously submit their comments and proposals, however, it is the employer who decides about their potential consideration. Potential bidders can obviously appeal, but the appeal procedure is also burdened with the risk of incurring sizeable costs, without allowing the appeal, hence bidders not always decide to appeal. They choose then another solution, including the risk costs in the bid price. The improper risk allocation results ultimately in higher prices in bids submitted by contractors, because they include the risk, which they cannot control, and which they have to accept.

The improper risks distribution also results from the fact that the contract patterns, developed frequently many years ago, are adopted uncritically and no experience-related changes are introduced in them, or such changes, which are aimed at increasing the effectiveness during the services provision. The justification of proceeding, in which the employer burdens the contractor with as wide as possible scope of risk, also results from the opinion, that the employer pays and thereby expects the contractor to perform the entrusted task comprehensively, and it is the contractor, who is the entity experienced in such projects implementation; their implementation is the subject of its business, hence it should cope well with the risks management and the issues, which can occur.

However, it is necessary to be aware, that the contractor will include the incurred risk in the payments for the provided services, will establish relevant financial reserves, or will adopt a solution, in which for a determined payment it will transfer the risk to an insurer, and in the situation, where the given risk materialises and a specific loss will occur, the insurer should pay the compensation offsetting this loss. So the insurance may be one of instruments to mitigate the risk effects by the policy holder, however, it definitely is not a solution, which minimises the risk related costs of the project under implementation. Also a situation can occur, where the contractor does not identify the risk, hence it is allocated to the party, which until its occurrence is not aware of that, which has no experience in controlling and minimising the probability of undesired event and its effects occurrence, which does not monitor circumstances affecting the risk and the undesired event occurrence. As a result, the effects in the form of losses, additional direct and indirect costs, can be significant, and frequently they also are not covered by insurers.

The lack of interest of operators providing the transport in actions aimed at increasing the achieved income is the basic drawback of contracts in the gross costs formula, the risk resulting from them is run by the transport organiser. This is not the best solution, considering that the income on the sales of services is one of most important sources of transport financing, and in addition it also reflects the adaptation of services to expectations of their users. Such situations may be changed modifying the division of risks in contracts; as a result we can refer to contracts of gross costs with motivation systems, or so-called hybrid contracts [19]. The change consists in sharing the risk of achieving an appropriate income on tickets sales by the organiser and transport provider. An assumption is then made that the entity serving a given line knows the current line utilisation, which allows it to propose changing timetables to the organiser, so that they would better match the passengers needs and thereby increase the demand for services. To a large extent a part of parameters related to the quality of provided services depends on the transport provider, and in particular on the driver; for example the cleanness of the vehicle's inside, timeliness, the way of vehicle driving, and possible provision of information, e.g. on the route of journey. Also the driver's involvement, if anticipated, in the ticket sales and inspections is important, and also the drivers cooperation with inspection services. This affects both the perception of public transport services and the amount of obtained income on tickets sales. The concept of contracts with motivation systems, or hybrid otherwise, is not new [20–24]; they enable obtaining benefits not only in the field of service effectiveness, but also in the social dimension [25]. The different effectiveness of such contracts is indicated (not always higher), which depends, for example, on the amount of motivating elements and the economic standing of the operator itself [26–29]. Such contracts are also used in regional public transport, and the basic objective of their introduction was the expectation of the increase in the transport volume [30]. Table 2 presents drawbacks and advantages of such a contract.

Table 2. Basic advantages and drawbacks of so-called hybrid contracts, i.e. with motivation elements

Type of contract concluded between organiser and operator	Basic advantages and drawbacks of gross costs, hybrid, and net costs contracts
Hybrid contract, i.e. contract with motivation elements	Advantages: - introduction of a solution, in which the operator can obtain additional remuneration in the case of proposals and co-participation in solutions resulting in the increased demand for services, hence in the increased effectiveness of the public transport system, - linking the operator's remuneration with the volume of transport and the income obtained on the sales of services Disadvantages: - demand for public transport services depends on at least a few factors, which frequently occur at the same time; hence it is difficult to determine the influence of actions taken by the operator on the demand increase or decline.ser, ananswers the passenger after the response receipt

Source: own study

The income on sales depends on the service prices and on the scope of provided services, and in this field the risks and their effects related to the taken decisions should be carried by these entities. However, the entities providing transport also affect the quality of services. The condition of vehicles, their cleanness, marking, reliability of service, timeliness, kindness and competence in the information provided to passengers, shown by drivers [31, 32] and traffic supervision services, and also the provision of any suggestions related to possible improvement in the transport offer, or just a possibility of introducing minor changes, all that can and contributes not only to the perception of the public transport, but also to decisions on the method of travelling. As a result, pursuing more efficient and effective systems of public transport, the contracts should contain provisions, pursuant to which the contractor's proposals and suggestions are considered, and thereby it has its share in the additionally obtained surplus income on tickets sales, above the planned level. The obtained additional income will then possibly supply not only the organiser, but will be divided between the organiser (public entity) and the operator [33]. At the same time it is necessary to consider a situation, in which the income on tickets sales will decrease below the planned level, how it was affected by passivity, whether the reason is related to the transport services provision, or, for example, it was influenced by events independent of that. If this is related to the quality of services, also the remuneration for the provided services can be partly reduced. The principles of bonuses awarding to operators for the maintaining or increasing the transport volume may differ, even up to a defined amount of additional remuneration for each transported passenger, above the number calculated for the first year of the contract implementation. However, in such systems the operator must have a subject-matter contribution during the services planning and development, because only then we can refer to the operator's involvement in the transport volume increasing [34]. Using a system of incentives and of financial bonuses awarding in the contracts, the organiser must so determine their level, that the bonus obtained by the operator would be lower than the effects achieved as a result of applying only incentives, hence, for example, comparing incomes and expenditures, so that the amounts paid to the operator for transporting an additional passenger would not be higher than the revenue obtained as a result of fare payment by the passenger. Also social issues of the services of urban public transport add to that, which causes that the mechanism of bonuses awarding must be more complex, it also should consider transport of persons entitled to free transport. Table 3 presents the risks location in hybrid contracts.

The application of solutions, in which the operator's remuneration is related to its involvement and actions for increased demand for transport services, requires establishing a relevant measurement system. However, it is difficult, despite equipping numerous vehicles of urban transport with many electronic systems, also those measuring the number of transported passengers, to link clearly the increase in demand change with specific actions of operators. Because diverse changes proceed all the time in transport systems, in their surroundings, and in the transport behaviour of residents. Prices of transport services, fuels, and parking charges change, which influences the scope of individual means of transport use, new jobs and places of residence originate in new locations, various transport investments are carried out, car parks in park & ride systems are started as well as new connections, lines, and many timetables change, adapting the offer to the needs. The continuity and multitude of changes occurring in cities cause, that even applying the methods of factor analysis it is difficult to make a precise assessment of the influence of various factors on e.g. the volume of transport. These are always approximate data, burdened with possible inaccuracies. However, this does not exclude a possibility of such solutions implementation and determination of reasons and effects based on the basic economic knowledge, related to the development of demand depending on the price, attractiveness of the service and substitutes. Traffic modelling and forecasting systems are another tool, in which models are developed, followed by traffic simulations and forecasts, depending on various factors and variables affecting just the traffic.

Irrespective of changes related to the risk of achieving income on tickets sales, it is also necessary to draw attention to the allocation of other risks. And so the public administration and entities organising the urban public transport should take the risk related to the amount of taxes paid by carriers/operators. In particular this should apply to taxes on the land, real property, and taxes imposed on the electricity and fuel, the more so that the tax income is partly the income of entities, which finance the public transport. In a similar way the amount of costs related to the fuel consumption is largely independent of the transport providers, and should burden the entity, which orders the services. The very fuel management depends on the transport providers, as well as motivation for economical driving and elimination of frauds, if any, while in the case of fuel purchases and supplies, if carried out in open competitive procedures, the operators cannot influence their prices, hence the risk of fuel price changes should be offset by the

Table 3. Risks allocation between the organiser and operator in hybrid contracts

Type of contract concluded between organiser and operator	Entity incurring specific risks	Risk description
Hybrid contract, i.e. gross costs contract with motivation elements	Organiser's risks	- risk of the planned size of demand, and ultimately of the income on sales of services, - risks related to possible irregularities in the provisions of or in the resolving of tariff regulations, - risks related to not ensured accessibility to the sales network, inspections, and effective recovery of additional charges, y of introducing changes and innovations
	Operator's risks	- risk of exceeding own costs and, as a result, impossibility of covering them in full from the payment for the provision of services obtained from the organiser, - risk of adopting adjustment formulae, which will not reflect the actual increase in
	Common risks	- risk of contract conclusion in accordance with the binding legal regulations, - risk of contract provisions ambiguity and related disputes during its execution,

Source: own study

employer of transport provision. At the development of remuneration systems for the provided services and the adjustment formulae it is necessary to predict such mechanisms of settlements, as to increase the remuneration by the increase in prices, and in justified cases by the increase in costs. Because it is also necessary to consider a situation, in which, for example, the introduction to operation of vehicles with additional equipment, e.g. air conditioning or various safety systems, most frequently results in the increased fuel consumption, and hence in costs increased above the value resulting from the price increase. In the contracts, in particular in the field of fuel and energy prices, a relatively quick mechanism of making adjustments and settlements should be predicted. Because if frequently the changes in tax burdens or due to other charges are made in annual periods, and they are known with certain advance, in addition they are not a major share of total costs, apart from payroll charges, then the changes in fuel or energy prices are a significant part of the incurred expenditures, and also they proceed in a way difficult for earlier planning. When making changes it is necessary to assume, that tax reductions, giving up selected charges, or fuel prices reduction, result in relevant reduction of remuneration for the transport provider.

Entities providing transport services should not incur the whole risk of remuneration increase and payroll related charges, because they do not fully control the increase in the payroll amounts. In numerous countries there are regulations on the minimum wage/salary, which means that changes are made there depending on the carried out policy, but also on the economic and social situation. The increase in the minimum wage/salary may be higher than inflation, in addition it causes a simultaneous general increase in the average pay level, the increased wages/salaries of part of employees (with the minimum pay) result in a general pay increase. Moreover, the increased remuneration of persons with the minimum pay results in the increase in the other employees remunerations, so as to maintain at least an appropriate pay structure, and the salaries differentiation between employees with higher qualifications and greater responsibility, and employees with lower qualifications. Hence the element related to the pay rise should be considered in adjustment formulae for contracts of the provision of transport services, not necessarily only in the form of adjustment by a general price increase index. A significant share of payroll costs in the total cost, and also the social nature of this cost element, and the influence of the pay level on the possibility of acquiring employees for work, causes a special importance of this cost component. Because the transport provider incurs the risk of securing resources indispensable for the services provision, i.e. employing a relevant number of drivers, and with appropriate qualifications, and of other persons necessary to run the business, with all the elements of requirements imposed on the drivers and affecting the safety. It should also carry the risk of effects related to the occurrence of unpredicted breaks in the transport services provision, for example due to collective disputes and potential breaks in the work.

The transport provider should carry risks related to the potential shortage of vehicles in good repair at a number and specification stipulated in the contract, and also to events, which could occur during the vehicles purchases, lengthening bidding procedures, missing appropriate protection of buyer's interest in the concluded contract, vehicles delivery with a delay, improper quality or parameters (e.g. fuel consumption) inconsistent with the declared ones, the quality of guarantee repairs, or problems with the spare parts delivering. In addition, the risk related to the impossibility of repairing vehicles in the planned time could be mentioned here. It should also carry the risks related to unfavourable weather conditions, because it is only it that can take actions to protect against them, unless these are extreme situations of force majeure nature.

The risks incurred by the transport services provider can include timetable changes, which for example result in the increased involvement of resources with simultaneous smaller amount of performed operational work, for which the remuneration is paid. The effects resulting from such changes could be considered in adjustment formulae, unless these are gross contracts with motivation mechanisms (hybrid), hence such, in which additional remuneration has been predicted on the obtained revenues on tickets sales. In such a case the timetable changes, due to which the carrier despite the same or even higher involvement of resources does not achieve adequate remuneration, are paid in accordance with the mechanism of additional remuneration due to the volume of transport, and thereby income on the sales of services.

In long-term contracts, in many cases the adopted formulae for contractor's remuneration do not take into account deteriorating traffic conditions in the city, which result in decreasing operational speed of vehicles, e.g. buses. The decline in operational speed means that in a given time a smaller number of vehicle-kilometres will be covered, hence a decrease in the drivers productivity and a lower remuneration on the performed operational work, at the same, and over time increasing, level of hourly rates of vehicle drivers. This risk should be transferred to the employer, hence a decrease in the operational speed should result in an increase in the rate paid for vehicle-kilometre, the amount of increase depends on the cost structure, i.e. the share of costs depending on the time of work or depending on the mileage. It may be mentioned, that in certain countries, apart from Poland, the remuneration dependent on the vehicles operation time is used, the rate is paid for a vehicle's operation hour. The solution, which should be considered, consists in making the payment dependent on two elements: vehicles mileage and operation time. It would then be not necessary to differentiate the rate paid depending on the traffic conditions, as it is the case for payments depending only on one factor, i.e. the number of travelled vehicle-kilometres. In such a situation the operator would not carry the risk of timetable changes, which result in incurring higher costs, for example related to the need to involve a larger number of vehicles, which will not necessarily cause an increased number of travelled vehicle-kilometres and hence obtaining a higher remuneration. A two-component remuneration, i.e. dependent on the mileage and time of work, takes into account the nature of costs incurred in transport. In addition, the situations of diversified rates per vehicle-kilometre calculated in one-component formulae, which result from different operational speeds, will be eliminated.

The organiser incurs the risk of proper conclusion of contracts with operators, in accordance with the stipulated procedures, so that there would be no reservations with respect to the operator's selection, and thereby proper spending of public funds. The contract provisions should be unambiguous themselves, they should not raise doubts and should be clearly controllable. The introduction of provisions, requirements, and duties, which the organiser would not be capable of controlling, or would not have appropriate inspection resources, will result, sooner or later, in their non-observation by the operator, which ultimately should result in the reduction or suspension of payments stipulated in the contract.

A common risk in long-term contracts should be a properly chosen formula for adjustment of remuneration rates paid by the organiser, so that it would consider changes of costs independent of the carrier, which it cannot affect.

The occurrence of phenomena of force majeure nature is another common risk, where it is proposed to specify this term in the contract so, that it would not be abused and would not result in an attempt to consider a force majeure phenomenon such events, which are not them. Also various unpredicted situations can occur during the contract duration, in relation to which the responsibility and rules of proceeding are not included in the contract. This partly results from a great diversity of events, which can occur during the contract, which means in practice the impossibility of determining all situations, which can occur, and in the case of their occurrence, the rules of proceeding, but also from the fact, that the occurrence of certain situations is unlikely, hence they have been neglected in the contract. In the case of selected events, e.g. being force majeure in nature, it is difficult to protect against them, and also their effects can differ. Hence only after their occurrence the necessary changes in the performance of the contract subject are assessed, and thereby amendments to the contract, so that in this situation, unfavourable for both or one of the parties to the contract, the optimum solution would be found, and thereby adverse effects and losses would not be increased.

5 Summary

The risk is an inherent element in entities management and operation, including also those, which organise and provide services in the urban public transport. The provision and organisation of the urban collective transport is related to many risks, in particular it is possible to mention the risks related to the adoption of specific organisational solutions, shaping the transport offer, decisions about service prices as well as tickets distribution and sales systems, procedures aimed at contractors selection, the concluded contracts, and settlement of contractors remuneration. The materialisation of risks and the occurrence of unwanted situations, which previously were only risks, can have various effects; one can mention the financial, legal, image, social, and environmental effects. Irrespective of the type, in most cases the negative effects are significant, which results from the universal nature of urban public transport services, the impact of services quality, prices, and public transport image on residents' decisions related to movements, and hence the division of transport tasks, the transport system efficiency, and also the effects for the city environment.

The financial results of risk materialisation mean the incurring of expenditures, which did not have to occur, and which could have been avoided, an increase in expenditures, and the reduction or not obtaining of the planned income. As a result, because of limited influence of market processes, this will mean the necessity of increasing the public financing of urban collective transport, or reducing the scope or the quality of provided services. The legal effects can have various dimensions, also an image-related and social one. They can be related to improperly made decisions, by unauthorised bodies or persons, errors in the procedures, or in the concluded contracts themselves. The image effects result both from the decisions taken on the scope and quality of transport services, prices - their price levels, and the scope of persons entitled to concessionary and free transport, the functioning of the service sales systems, but also of the system for informing about services, changes, and consideration of applications and claims. This affects not only the public transport itself, but also the level of city perception, friendliness, and the conditions of life in the city, carrying out business, and competing with other cities.

In big cities and conurbations, in the systems of urban public transport regulated and financed from public funds, contracts in the gross costs formula are applied. This results from many advantages of the division of tasks and risks between the organiser and operators of urban public transport adopted in them. In the organisation of collective transport, which comprises the organising entity and gross costs contracts with operators, it is possible to control the spent public funds, as well as the functioning of many operators without competition on routes and during the services provision, coordination of various operators' timetables, a uniform and integrated tariff, and common transport information. In such contracts the tasks and risks are properly allocated to the organiser and operators carrying out transport. The organiser, defining the scope of transport tasks and the tariff, incurs the risk of declining demand for services, and the risk related to the difficulty in ensuring funds for the public transport at the required amount, and also such, which results from the incurred long-term liabilities. In turn, the operators incur the risk of ensuring resources necessary to provide transport services, and not exceeding at the same time the level of costs, which will be financed by the organiser.

Despite a basically positive assessment and acceptability of such a solution for big cities and conurbations, it is possible to show a need, and even a challenge related to the modification. This results from the necessity of actions aimed at improvement in the effectiveness of the urban public transport, stopping the decreases or increasing the number of transported passengers, or reduction of the dynamics of growth, or stabilisation of the amount of public financing. The contracts in the gross costs formula cause that operators focus primarily on costs, maintaining the costs level at the amount of payment received from the organiser on the sales of services. The transport organiser, and not the passengers, is the employer and the entity, which approves the provided services. The organiser is obviously carrying out relevant marketing studies, it knows the market and transport needs, however, the vehicle drivers and other services of operators are frequently closest to the service users. Unfortunately, in such a contract model the operator and its employees are not motivated to cooperate with the organiser, in actions aimed at changes in the offer, which will result in the improved attractiveness of the offer, and hence the competitiveness of the urban transport services will increase in relation to other methods of travelling. They are not interested in the reduction of the transport volume in places, where they are no longer justified, and sometimes the reduction is not favourable to the operator from the point of view of transport potential utilisation. A lower demand for services outside the rush hours may be an example, in particular between the morning and afternoon rush hours; the operators are not interested in lines with such timetables, because they cause lower utilisation of vehicles, and also an inconvenient system of working hours for the drivers. The operators involvement in the process of timetables improvement, and thereby of the transport offer, tickets sales in vehicles, e.g. during night hours and on holidays, ticket inspections carried out by the drivers, and cooperation with the organiser's inspection services, will affect the amount of income on tickets sales, and thereby should influence the scope and quality of services of the urban collective transport.

The structure of operators remuneration may be another drawback of contracts in the gross costs formula, in particular in the situation of dynamic changes of fuel and energy prices, and other cost items. The operators do not have any influence on a part of costs incurred by them, like taxes, for other they have a rather limited impact. The varying costs, e.g. increasing amounts of fiscal burdens or of market prices, should be compensated to operators by the use of price forecasts, and later on by the adjustment of the remuneration itself; it is not the very amount that matters, but also that the adjustment would not be substantially delayed in relation to the moments, when it is necessary to incur higher expenditures. For example, the annual or quarterly settlements can mean a pretty significant shift in time and additional costs for operators.

From the point of view of improvement in the effectiveness of provided services and the utilisation of resources it is difficult to assess positively solutions, in which the amount of remuneration covers the justified costs and ensures a reasonable profit. Because the term of justified costs is very broad, it does not replace the market mechanisms and competition. The remuneration up to the amount of incurred costs reduces the motivations for actions aimed at their reduction, it rather releases actions aimed at their justification and presentation of a given service benefits. Higher effectiveness/lower costs appear in solutions, in which contracts are concluded in competitive procedures. Unfortunately, the possibility of choosing, and hence giving up comparative procedures, is not replaced with other mechanisms in this field; it is frequently difficult, due to the incomparability of cities and urban transport systems, to use a comparative method for making assessments. Also the conduct of public control, or participation in authorities of entities carrying out transport, will not replace costs cutting mechanisms, which are characteristic of competitive markets. Because actions in favour of the public transport sector concentrate then on the pursuit of ensuring appropriate funds in public budgets, and not on the efforts to improve the effectiveness. In many cases it happens this way at the cost of increasing the public expenditure in budgets of various levels, to the amount exceeding the obtained income.

Systems of contracts in the urban public transport, despite many years of this service existence and its public financing, have not been standardised yet, as it is the case for contracts of construction works. In numerous cases these are individual solutions in cities, not denying the cities rights to decide, or the principle of subsidiarity. So it is necessary to consider making an attempt to create a certain formula, the basis of such contracts, to facilitate their conclusion, to avoid part of drawbacks, and also to establish certain standards of the urban public transport service in supra-national systems.

Also a properly formulated contract, in particular such, in the contents of which risks related to its implementation are visible, they are allocated to the parties according to the principle, that the risk should be taken by the entity, which can affect it, monitor it, and prevent its occurrence, and also which fastest obtains the information on its materialisation, is helpful in identification and minimisation of possible risks effects. The knowledge about the existing risks and the implementation of their management processes allow for actions aimed at the reduction of occurrence probability of an unwanted negative event, and if it occurs, at reduction of its effects. The limitation of effects results from acquiring the information about the event occurrence possibly quick, and from taking previously prepared actions, aimed at reduction of negative effects. It is necessary to emphasise that this refers to the risk, which is situated in a given entity.

References

1. Regulation (EC) No 1370/2007 of the European Parliament and of the Council of 23 October 2007 on public passenger transport services by rail and by road and repealing Council Regulations (EEC) Nos 1191/69 and 1107/70 (2007). https://eur-lex.europa.eu/legal-content/EN/TXT/?uri=celex%3A32007R1370

2. European Commission, Study on the Implementation of the European Regulation (EC) No 1370/2007 on public passenger transport services by rail and by road of 23 October 2007 (2007). https://transport.ec.europa.eu/transport-themes/public-service-obligations/studies_en
3. Gautier, A., Yvrande-Billon, A.: Contract renewal as an incentive device. An application to the French urban public transport sector. Rev. Econ. Inst. **4**(1), Article 2 (2013)
4. Currie, G.: Deregulation, Franchising, Outsourcing and Corporatisation in Local Public Transport: International Experience, International Transport Forum Discussion Papers 2016/10. OECD Publishing (2016)
5. Smith, A., Nash, C.: Procurement of public transport: contractual regimes. In: Vickerman, R. (ed.) International Encyclopedia of Transportation, pp. 308–314. Elsevier (2021). ISBN 9780081026724. https://doi.org/10.1016/B978-0-08-102671-7.10511-1
6. Beck, A.: Experiences with Competitive Tendering of Bus Services in Germany. Transp. Rev. **31**(3), 313–339 (2021). https://doi.org/10.1080/01441647.2010.511299
7. Organization for Economic Cooperation and Development: Privatization and Regulation of Urban Transit Systems, Transport Research Centre. OECD International Transport Forum (2008). https://doi.org/10.1787/9789282102008-en
8. Steer Davies Gleave, Study on economic and financial effects of the implementation of Regulation 1370/2007 on public passenger transport services, prepared for European Commission, Final Report February (2016)
9. Hensher, D., Gwilliam, K., Burton, M., Smith, N., van de Velde, D., Fridstrom, L.: The ideal contract roundtable. Res. Transp. Econ. **22**(1), 188–194 (2008). https://doi.org/10.1016/j.retrec.2008.05.016
10. Russell Bernard, H.: Sociale Research Methods: Qualitative and Quantitaive Approaches. SAGE Publications, Inc., Los Angeles, London, New Delhi, Singapore, Washington D.C. (2013). ISBN 1412978548, 9781412978545
11. Mintzberg, H.: An emerging strategy of 'direct' research. Adm. Sci. Q. **24**(4), 582 (1979). https://doi.org/10.2307/2392364
12. Shah, S.K., Corley, K.G.: Building better theory by bridging the quantitative? Qualitative divide. J. Manage. Stud. **43**(8), 1821–1835 (2006). https://doi.org/10.1111/j.1467-6486.2006.00662.x
13. Czernek, K.: Qualitative research in economic sciences – usefulness and challenges. The case of the research of small and medium sized enterprises cooperation in a tourist region. Problemy Zarządzania **12**(3(47)), 164 (2014). https://doi.org/10.7172/1644-9584.47.9
14. Hoang, H., Antoncic, B.: Network-based research in entrepreneurship: a critical review. J. Bus. Ventur. **18**(2), 165–187 (2003). https://doi.org/10.1016/s0883-9026(02)00081-2
15. Lechner, C., Dowling, M.: Firm networks: external relationships as sources for the growth and competitiveness entrepreneurial firms. Entrep. Reg. Dev. **15**, 1–26 (2003). https://doi.org/10.1080/08985620210159220
16. O'Donnell, A., Gilmore, A., Cummins, D., Carson, D.: The network construct in entrepreneurship research: a review and critique. Manag. Decis. **39**(9), 749–760 (2001). https://doi.org/10.1108/EUM0000000006220
17. Babbie, E.: Podstawy badań społecznych. Wydawnictwo Naukowe PWN, Warszawa (2008). ISBN 978-83-01-21274-2
18. Bitetkine, A.: Prospective Case Study Design. Qualitative Method for Detuctive Theory Testing, Organizational Research Method, Copyright 2007 by SAGE Publications (2007). https://www.academia.edu/382335/Prospective_Case_Study_Design
19. Van de Velde, D., Beck A., van Elburg J.C., Terschüren K.H.: Contracting in urban public transport. ResearchGate (2008). https://www.researchgate.net/publication/259900353_Contracting_in_urban_public_transport

20. Hensher, D.A., Houghton, E.: Performance-based quality contracts for the bus sector: delivering social and commercial value for money. Transp. Res. Part B Methodol. **38**(2), 123–146 (2004). https://doi.org/10.1016/s0191-2615(03)00004-3ISSN1440-3501
21. Hensher, D.A., Mulley, C., Smith, N.: Towards a simplified performance-linked value for money model as a reference point for bus contract payments. Res. Transp. Econ. **39**(1), 232–238 (2013). https://doi.org/10.1016/j.retrec.2012.06.018
22. Danielson, H., Andersson, H., Wretstrand, A.: Superincentive contracts: a study of the VBP contract models in Stockholm–draft version. (p. 1). (K2 Working Paper). K2 - Nationellt kunskapscentrum för kollektivtrafik (2016). https://lup.lub.lu.se/search/ws/files/7270689/superincentive_contracts_2016_5.pdf
23. Gómez-Lobo, A., Briones, J.: Incentives in bus concession contracts: a review of several experiences in Latin America. Transp. Rev. **34**(2), 246–265 (2014). https://doi.org/10.1080/01441647.2014.895451
24. Vigren, A.: Incentives in public transport contracts. In: International Encyclopedia of Transportation, pp. 332–336. Elsevier (2021). https://doi.org/10.1016/b978-0-08-102671-7.10063-6
25. Hensher, D.A., Stanley, J.: Performance-based quality contracts in bus service provision. Transp. Res. Part A Policy Pract. **37**(6), 519–538 (2003). https://doi.org/10.1016/s0965-8564(03)00006-5. ISSN 0965-8564
26. Gagnepain, P., Ivaldi, M.: Contract Efficiency in Public Transport Services (version 18.10.2019) (2019). https://www.parisschoolofeconomics.eu/docs/gagnepain-philippe/ch0000_un-102_gagnepain-18-oct-2019(1).pdf
27. Sheng, D., Meng, Q., Li, Z.: Optimal quality incentive scheme design in contracting out public bus services. Transp. Res. Part C Emerg. Technol. **133**, 103427 (2021). ISSN 0968-090X
28. Vigren, A., Pyddoke, R.: The impact on bus ridership of passenger incentive contracts in public transport. Transp. Res. Part A Policy Pract. **135**, 144–159 (2020). https://doi.org/10.1016/j.tra.2020.03.003. ISSN 0965-8564
29. Pyddoke, R., Lindgren, H.: Outcomes from new contracts with "strong" incentives for increasing ridership in bus transport in Stockholm. Res. Transp. Econ. **69**, 197–206 (2018). https://doi.org/10.1016/j.retrec.2018.07.022. ISSN 0739-8859
30. Alexandersson, G., Hultén, S., Jardón, J.J.: Hybrid markets in public transport – contract design, performance and conflicts. Res. Transp. Econ. **83** (2020). https://doi.org/10.1016/j.retrec.2020.100897
31. Chaparro, A., Galilea, P., Muñoz, J.C., Poblete, J.: Application of an incentive for bus drivers to achieve an improvement in the quality of service. Res. Transp. Econ. **83**, 100908 (2020). ISSN 0739-8859
32. Tiznado, I., Galilea, P., Delgado, F., Niehaus, M.: Incentive schemes for bus drivers: the case of the public transit system in Santiago. Chile. Res. Transp. Econ. **48**, 77–83 (2014). https://doi.org/10.1016/j.retrec.2014.09.034. ISSN 0739-8859
33. Walters, J.: Potential cost implications of contracting risks – the views of bus operators in South Africa. Res. Transp. Econ. **69**, 235–244 (2018). https://doi.org/10.1016/j.retrec.2018.03.009
34. Piechucka, J.: Cost Efficiency and Endogenous Regulatory Choices: Evidence from the Transport Industry in France, Discussion Papers of DIW Berlin 1851, DIW Berlin, German Institute for Economic Research (2020). https://www.diw.de/documents/publikationen/73/diw_01.c.725610.de/dp1851.pdf

Identification of Risks Related to Passenger Exchange Service in the Water Tram Stop Area on the Example of the Water Tram of the Wrocław Agglomeration

Emilia T. Skupień(✉), Agnieszka A. Tubis, and Mateusz Rydlewski

Wroclaw University of Science and Technology, Wroclaw, Poland
{emilia.skupien,agnieszka.tubis,mateusz.rydlewski}@pwr.edu.pl

Abstract. The paper presents aspects of risks related to the infrastructure of the water tram stop area on the example of the water tram of the Wrocław Agglomeration. The research area was limited to the passenger exchange service process. To identify the risks, the HFACS (Human Factors Analysis and Classification System) analysis framework was used. Then the city standards for the requirements of the tram and bus stops in Wrocław and the standards for the infrastructure of the stops of water vessels were analysed. Based on that, the minimum requirements must be met by water tram stops located in the urban agglomeration. These standards are a baseline that must be completed in the design and modification of the current water tram stop.

Keywords: water tram · risk · passenger exchange · water tram stop area

1 Introduction

Along with the development of civilisation, the necessity for sustainable development is noticeable in every area of life. This also applies to sustainable urban mobility. It is currently one of the most important trends in the development of European cities.

The need for a comprehensive approach to transport and environmental issues in urbanised areas was emphasised, among others, in [20]. This study highlights that the most important challenges emerging at the interface between transport processes and the city include [20]:

- transport behaviour unfavourable from the point of view of sustainable development, resulting in the division of transport tasks with the dominant role of the passenger car;
- the progressive spread of urban areas, resulting in a greater dependence on the passenger car and an increase in the distance between the source and destination of traffic;
- impact of the transport sector on climate change;
- air pollution in cities;

G. Sierpiński et al. (Eds.): TSTP 2023, LNNS 907, pp. 34–45, 2024.
https://doi.org/10.1007/978-3-031-53181-1_3

- traffic safety in the city zone;
- negative effects of noise generated by intensive road traffic;
- strong dependence of all means of transport in the city on fossil fuels.

In connection with the challenges defined in this way, it is right to focus primarily on aspects related to the functioning of public transport, pedestrian and bicycle traffic, door-to-door mobility of residents, safety, urban road transport, and city logistics. For this reason, as part of the development of sustainable urban mobility, decision makers focus on the integration of means of transport and increasing the degree of use of existing transport infrastructure and transport services.

To become more sustainable, urban transport needs, in addition to the necessary infrastructure, also diversification of the possibilities of moving within the urban agglomeration. The developed and offered solutions for moving within the city should therefore be aimed at increasing free-flow movement, efficiency, and ecology in urban traffic. Therefore, it is reasonable to transfer part of it from the city roads to the river.

At the same time, there is increasing ecological awareness in society and a desire for changes aimed at reducing pollution that lowers air quality in cities.

The concept of creating the Water Tram of the Wrocław Agglomeration (TWAW) proposed in Wrocław is part of the National Urban Policy 2030 [8]. The main objective of the project is to reduce the share of cars (individual transport) in the total share of all transport in the Wrocław Agglomeration. It is a solution based on the natural waterway and a return to its original importance for the development of large cities. It will allow one to supplement the existing urban transport system with an ecological solution that can relieve the current public transport system not only in the area of the city centre, but also in agglomeration connections.

TWAW is an exemplary example of turning the city "face to the river" and its use in balancing the mobility of residents and travellers coming to the city. It is also a response to the city model currently promoted by the European Union, in which transport is to be effective, resource-efficient, ecological, and low-emission. Therefore, the water tram fits the current vision of sustainable urban mobility.

An important element of the created solution is to ensure its accessibility and safety for various groups of passengers, including people with disabilities and special needs (e.g. elderly people and children travelling in prams). The appropriate adaptation of stop points to the needs of people with reduced mobility may be an important element in popularising this means of transport in urban travel.

Stops of water tram combine the risks of stops as they are and the aspects of water transport. Therefore, it is important to refer to standards and adverse events in both fields.

Stops imply the interaction between passengers and a vehicle. This may lead to adverse events. It was the subject of several publications, among others: [1, 3].

Also accessibility for people with movement problems (not only elderly or invalid people, but also passengers caring big language or with a baby trolley) is a subject of interest [6]. The [24] refers to accessibility of public transport stops and its specifics, showing the obstacles and standards that allow them to be overcome. Accessibility is also important in increasing the sustainability of public transport [4, 21, 22].

The risks related to inland waterway transport were under investigation due to the possibility of the ship sinking, collision, grounding, man falling overboard, and others directly related to the presence of the ship on the water. It was described in research as: [7, 19, 23].

Based on the experience and knowledge of the authors and the invited experts, the identification of risks related to the passenger exchange service in the water tram stop area on the example of the Water Tram of the Wrocław Agglomeration was proceeded.

2 Methodology

Research on the risk associated with the exchange of passengers at water tram stops was part of the research on developing standards for stops for the Wrocław Water Tram. For this reason, the research procedure included three stages presented in Fig. 1.

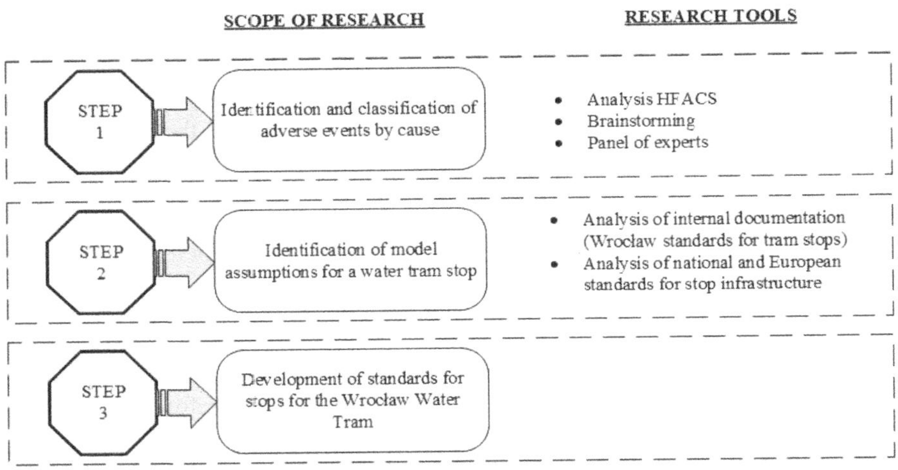

Fig. 1. Stages of the research procedure

The first stage of the research was to determine possible adverse events that may occur during the exchange of passengers at the city water tram stops. Identifying the risk associated with the exchange of passengers is critical from the point of view of developing guidelines for the Wrocław Water Tram stop standard. The developed standards should ensure the required level of safety at every stage of the journey. Passenger exchange is critical in this process due to the numerous adverse events that can occur during its course.

The team of experts participated in the process of identifying adverse events due to the lack of historical data that could be the basis for quantitative analysis. Representatives from the city responsible for the stop infrastructure, people with knowledge and experience in the field of water transport, and an expert in the risk area were invited to the team of experts. The HFACS (Human Factors Analysis and Classification System) analysis framework was used to identify the risk. This method develops the concept of

"Swiss cheese" [5], the slices of which represent protective barriers for the tested system. Some holes in these slices are related to unintentional flaws in the system mapped in each slice. These holes may affect the possibility of an adverse event when the holes of each barrier align in a straight line [16]. The HFACS analysis refined the notion of "holes" occurring in successive cheese slices, assigning four areas of required research regarding the tested system/process. They are [18]: (1) Unsafe Acts, (2) Preconditions of Unsafe Acts, (3) Unsafe Supervision, and (4) Organisation Influence. These categories consider the causes of human errors due to individual process participants and constraints at the organisational level. Due to this, it is possible to investigate the causes of accidents at various levels of complexity [28]. Due to the scope of the assessment, in the analyses described in this article, the HFACS method was limited to two slices: preconditions of unsafe acts and unsafe acts.

The next stage of the research was the development of model assumptions regarding the stop infrastructure created. For this purpose, it was necessary to recognise city standards regarding requirements for tram and bus stops in Wrocław and national and European standards regarding conditions for stop infrastructure.

Based on the analysis of the collected standards and requirements for the stop infrastructure and the results of the risk assessment, minimum requirements were developed that must be met by the water tram stops located in the urban agglomeration. These standards are a baseline that must be completed in the design and modification of the current water tram stop.

3 Identification of Risks Related to Passenger Exchange Service in the Water Tram Stop Area on the Example of the Water Tram of the Wrocław Agglomeration

Identifying potential adverse events that may occur during passenger exchange is necessary to ensure a high level of service for passengers using the water bus. Qualitative methods were used to identify the risk due to the lack of historical data on which it would be possible to perform quantitative analyses. The research procedure is presented in Fig. 1. Following the assumptions adopted, the experts had to identify adverse events caused by two groups of reasons: (1) precondition for unsafe acts and (2) unsafe acts.

The precondition for unsafe acts applies to all factors whose occurrence may be the source of identified adverse events. According to the HFACS concept, we divide these factors into three groups [27]: (1) Environmental Factors, (2) Condition of Operators, and (3) Personnel Factors. Since the exchange of passengers at the water tram stop takes place under the supervision of the crew, but without ensuring complete control over the behaviour of passengers, (2) the group of risk factors was modified and changed to "Condition of Passengers". The results of the identification procedure are presented in Table 1.

The second source of adverse events is unsafe acts. According to the HFACS concept, the factors in this area are divided into two groups [27]: (1) Errors and (2) Violations. Since the source of an undesirable event may be errors and violations on the part of

Table 1. Adverse events in the area of 'Precondition for unsafe acts'.

Group	Symbol of an adverse event	Description of the adverse event
Environmental Factors	EF-PUA1	Waves on the river hinder the exchange of passengers
	EF-PUA2	Atmospheric phenomena limiting visibility (rain, snow, fog)
	EF-PUA3	Precipitation causes a slippery surface
	EF-PUA4	Damaged gangway elements
Personel Factor	PF-PUA5	Limited availability of personnel handling the passenger exchange process
	PF-PUA6	Lack of knowledge of the operator regarding the applicable passenger handling procedures
Condition of Passengers	CP-PUA7	Passenger under the influence of drugs or alcohol
	CP-PUA8	Panic among passengers
	CP-PUA 9	Lack of knowledge of passengers on precautions
	CP-PUA10	Excessive haste of passengers when exchanging

operators servicing the exchange of passengers and the passengers themselves, the identification of adverse events was divided into two groups of participants in the passenger exchange process. The results of the procedure conducted are presented in Table 2.

Table 2. Adverse events in the area of 'Unsafe acts'.

Group	Symbol of an adverse event	Description of the adverse event
Errors of Passengers	EP-UA1	Invalid gangway entry
	EP-UA2	A sudden change of decision regarding the direction of movement
	EP-UA3	Running in/out to/from the vehicle
	EP-UA4	Failure to comply with the rules of movement within the stop
	EP-UA5	Pushing passengers around the gangway

(continued)

Table 2. (*continued*)

Group	Symbol of an adverse event	Description of the adverse event
	EP-UA6	Attempting to jump into the vehicle during manoeuvres
Errors of Crew	EC-UA7	Too early release of passengers for replacement
	EC-UA8	Incorrect mooring of the vehicle
	EC-UA9	Incorrect assessment of the conditions around the stop (slippery surface, visibility, etc.)
	EC-UA10	Incorrect gangway protection
Violations of Crew	VC-UA11	Failure to check the required security features
	VC-UA12	Lack of supervision by the operator on the exchange of passengers

The identified undesirable events became the basis for preparing recommendations for creating passenger service procedures, operator training, and supplementing the minimum standard for the Wrocław Water Tram stop infrastructure.

4 Analysis of Legal Requirements with Respect to the Water Tram Stop Infrastructure

The infrastructure of the stop located in the area of the tram stop has a direct impact on the risks that can arise during the exchange of passengers. Although individual stops may differ from each other due to organisational and legal issues, certain requirements are collected in legislative documents, recommendations, and standards.

Guidelines describing the requirements for the water tram stop infrastructure can be found in a number of documents which, for the purposes of this publication, have been divided into two categories: 1) established by the city authorities where the public transport stop is located - here: Wrocław (taking into account higher-level guidelines, supplemented with local recommendations), 2) concerning paid carriage of passengers by inland water vessels.

The requirements indicated in this way have been collected taking into account their impact on the risk of adverse events during passenger exchange and presented in the form of minimum requirements that absolutely must be met by water tram stops in Wrocław due to passenger safety.

4.1 Analysis of Standards Regarding the Stop Infrastructure Applicable to the City of Wrocław

The issues related to the standards for stop infrastructure as part of the collective transport systems of trams and buses for Wrocław are mainly described in the following documents:

- Wrocław standards for the design of pedestrian-friendly urban spaces [25],
- Wrocław standards of accessibility of urban spaces [26],
- Catalog of street furniture [2],
- Cultural Park 'Old Town' [17].

Wroclaw stops can be divided into several groups, for which individual requirements may differ, but for the purposes of this study, basic elements consistent for all types of stops are crucial. In all cases, the recommended minimum width of the stop platforms is 3.5 m. Most stops are equipped with permanent elements of infrastructure, such as:

- Stop shelter - usually three-span with wide glass side walls, on which there are special stickers or spot sandblasting of the glass to prevent the walls from being invisible to birds or people. However, it should be remembered that the key element in terms of visibility is to ensure it from the approach side of the vehicle that serves the stop. From the point of view of the water tram, the view of the mooring unit will be from the direction of the entrance to the shelter.
- Bench and trash can.
- Dynamic passenger information board.
- Fencing off the platform space with a railing. It is used only for potential threats, depends on the specific surroundings of the stop, and is implemented in order to prevent direct entry of the passenger, e.g. on the bicycle path. In such circumstances, fencing with one or two bays of the railing is used directly at the bus shelter.

Each stop has its own stop post. There are 3 standard types and one for marinas. In addition to the sign informing about the stop required by law, there is also a schedule and information about changes in routes and ticket prices.

An increasingly common solution is the use of facilities for people with reduced mobility, which include warning stripes at the edge of the platform, guide strips, or suggesting waiting places for people with reduced mobility. Internal standards stipulate the use of guide and warning stripes with a width of 30 to 40 cm. Warning stripes should be 50 cm from the edge of the platform. The guiding lanes should be located on the axis of the lane free from infrastructural obstacles, 1.8 m wide. The guiding lanes usually lead from pedestrian crossings to the warning strip at the edge of the platform or the suggested waiting area for people with reduced mobility. It is recommended that they be led along the entire length of the platforms. Suggested waiting places for people with disabilities have a minimum size resulting from the standards 0.9 m × 1.0 m.

According to the regulations resulting from the "Stare Miasto" Cultural Park, the piers of passenger ships covered by these regulations are marked with information boards in accordance with the pattern, mounted on appropriate posts, in accordance with the catalogue of city furniture. At the stop of passenger ships, outside the time of their operation, it is not allowed to place objects other than those required by the regulations relating to the navigation of ships. On the premises of the passenger ship stop during its operation, it is allowed to place only:

- a desk for selling tickets with dimensions not larger than 60 cm wide, 60 cm deep, 100 cm high,
- 1 chair,
- 1 umbrella,
- items required by the laws relating to shipping and ships.

The stop equipment standards are key elements for the functionality and safety of the facility. Stop areas can be enriched with elements of vegetation that affect aesthetics and rainwater management. In this case, the minimum requirements specified in the standards must be ensured so that the green elements do not limit the functionality or deteriorate the safety of users.

4.2 Analysis of Standards Regarding the Inland Waterway Transport Infrastructure

Due to significant functional differences between traditional collective transport and a water means of transport, a water tram, in the further part of the study the requirements for water tram stops have been collected and presented.

Water tram stops should be characterised by geometric parameters corresponding in length to the parameters of vessels operating the water tram system. Along the length of the stop platform, accessibility to the entrance (or entrances or a system of entrances and exits) and space for equipment related to mooring should be ensured. The length of the stop should be expressed as a minimum range that must be met in order to realise a safe passenger exchange. The width of the stop platform should be adapted to the size of the passenger flow and, similarly to its length, it should be determined in the standards as a minimum. The minimum width of the platform is assumed to be 4 m. This is 0.5 m more than expected for standard stops in Wrocław. This is due to the fact that there is water at the stop line and the increase in width is dictated by safety considerations.

Additional requirements for water tram stops, resulting from the specificity of inland waterway transport, result from local regulations of the Director of the Inland Navigation Office and the opinions of the Inland Navigation Office and the Regional Water Management Board issued in the process of obtaining water-legal permits for the use of a specific place for the purpose of passenger exchange. General safety recommendations assume the following:

- gangways, i.e. devices used to board the vessel, must comply with PN-EN 14206:2006 [12]; in particular: they should have handrails on both sides, the minimum width of the gangway is 60 cm, and for the movement of people with reduced mobility: 90 cm; the length of the device should be adapted to water fluctuations up to 50 cm; they must be secured against slipping (have rungs made of slats or flat bars);
- barriers (outside the mooring area, which must be free of obstacles), must comply with PN-EN 711:2016-08 [15]; in particular: muntin may only be directed vertically, in a way that prevents climbing on the railings;

- floating pontoons (if needed) must meet the construction requirements (PN-EN 14504:2019-05 [14]) and must be attached to the shore in a way that allows levelling changes in water level up to 50 cm; they must be equipped with deck railings, in accordance with PN-EN 711:2016-08 [15], which will separate the spaces outside the pontoon entrance and the unit entry area from the water, in practice, the railings are set from the land side outside the gangway and from the sides; the side from the axis of the river leaving it free of obstacles;
- rescue equipment, including lifebuoys and boathooks, are on board the vessels. For this reason, it is not necessary to place them on floating pontoons, provided that the entrance to them is closed when the vessel is not moored there; in the case of pontoons accessible to passengers in the absence of the vessel, the lifebuoy should be in accordance with PN-EN 14144 [10], have a 30m floating line, and be placed on a holder in accordance with PN-EN 14145 [11];
- the distance between the mooring devices should not be greater than the length of the units moored to them, so as to enable attachment to two elements: from the bow and stern of the unit; mooring devices must comply with EN 13573 [9]; In the place of passenger service, these should be bollards or bollards in niches (PN-EN 14503: 2006 [13]);
- the minimum free working space for independent mooring devices is 1.0 m;
- on the shore there should be information boards containing a diagram of rescue and resuscitation of drowned persons and emergency numbers (PN-EN 14503: 2006 [13]);
- the shore of the ship should be marked with a sign;
- in the vicinity of ships moored on the shore, marina, or ships on which goods are transhipped or passengers are transferred, it is prohibited to generate waves.

5 Identification of Minimum Requirements for Infrastructure of Water Tram Stops

Based on the presented risks and standards that must be met by water tram stops in Wrocław, identification of the minimum requirements for infrastructure of water tram stops was made.

Infrastructure related to the water tram passenger exchange area should include:

- A standard three-span shelter for city stops with a bench in its central part.
- Wastebasket.
- Passenger harbour post with necessary information on the functioning of the water tram system - from the city furniture catalogue.
- Designated suggested waiting area for people who are blind or have reduced mobility and need assistance boarding.
- Guide belts for the blind.
- An information board with a diagram of the rescue and reanimation of drowned people and emergency numbers, as well as the regulations for using the stop.
- Mooring bollards.
- Mooring dolphin - when it is necessary to permanently moor a floating pontoon, which is part of the stop.

In addition, the stop should be fully illuminated, and if there is a ramp leading to the stop, it should be made of nonslip material or covered with an appropriate coating. Access to the stop should be adjusted so as to ensure comfortable movement on wheelchairs and it should be possible to equip it with belts for the blind, leading to the edge of the platform. For slopes above 5%, a handrail should be used at the ramp to enable catching by pedestrians coming to and leaving the stop.

Along the edge of the platform, along its entire width from the shore side, there should be a contrasting-coloured (yellow or white) "stop" warning plate, 40 cm wide. The warning strip should be located at least 50 cm from the edge of the stop platform.

Access to the stop should be accessible to people with reduced mobility; therefore, it is not recommended to use stairs as the only way to get to the stop platform.

6 Conclusions

The development of sustainable use of various modes of transport for passengers broadens travel options, increases the reliability of the system, and affects the conscious use of resources. However, with this development come risks as well. Some are identical to co-related to adverse events in land transport (e.g. accessibility in case of changes in height, or the possibility of slipping), and some are different (e.g. falling into water). Similarly, adverse events may have different effects if they occur (e.g. entering the road in an unauthorised place vs. falling off the platform). For this reason, it was important to identify the risks that may occur during the exchange of passengers at water tram stops, then confront them with the standards for the stop infrastructure and present the minimum requirements that such places should meet.

As a result of the research procedure, it was indicated that the safety requirements for inland waterway passenger transport are strict and clear. The legal standards regarding infrastructure cover all mentioned adverse events. Only suggested enlargement may concern the crew training and the passengers awareness.

References

1. Sustainable Urban Transport Plans. Preparatory Document in relation to the follow-up of the Thematic Strategy on the Urban Environment. European Commission, Office for Official Publications of the European Communities, Luksemburg (2007). https://www.google.com/url?sa=t&rct=j&q=&esrc=s&source=web&cd=&cad=rja&uact=8&ved=2ahUKEwjWmdnilLmAAxVlJxAIHR0bAxQQFnoECA4QAQ&url=https%3A%2F%2Fwww.gtkp.com%2Fdocument%2Fsustainable-urban-transport-plans%2F&usg=AOvVaw3lhVRJzP5DNZBCnbU4PdEp&opi=89978449. Accessed 5 May 2023
2. National Urban Policy 2030, Krajowa Polityka Miejska 2030 – Zapraszamy miasta do debaty (2021). https://www.miasta.pl/aktualnosci/krajowa-polityka-miejska-2030-zapraszamy-miasta-do-debaty. Accessed 1 May 2023
3. Abenoza, R.F., Ceccato, V., Susilo, Y.O., Cats, O.: Individual, travel and bus stop characteristics influencing traveller safety perceptions. Transp. Res. Rec. **2672**, 19–28 (2018). https://doi.org/10.1177/0361198118758677
4. Catalog of street furniture. https://bip.um.wroc.pl/artykul/100/3210/katalog-mebli-miejskich. Accessed 5 May 2023

5. Friman, M., Lättman, K., Olsson, L.E.: Public transport quality, safety, and perceived accessibility. Sustainability **12**, 3563 (2020). https://doi.org/10.3390/su12093563
6. Tubis, A.A., Skupień, E.T., Rydlewski, M.: Method of assessing bus stops safety based on three groups of criteria. Sustainability **13**, 8275 (2021). https://doi.org/10.3390/su13158275
7. Corazza, M., Favaretto, N.: A methodology to evaluate accessibility to bus stops as a contribution to improve sustainability in urban mobility. Sustainability **11**, 803 (2019). https://doi.org/10.3390/su11030803
8. Transport of London. Accessible Bus Stop Design Guidance; Public Transport Team: London, UK (2017). https://content.tfl.gov.uk/bus-stop-design-guidance.pdf. Accessed 2 May 2023
9. Tubis, A., Ratajczak-Szponik, N., Rydlewski, M., Skupień, E.T.: Identification of barriers to ensuring accessibility standards for tram stops to the needs of people with disabilities. In: Sierpiń, G., Masoumi, H., Macioszek, E. (eds.) Challenges and Solutions for Present Transport Systems. TSTP 2022. LNNS, vol. 609, pp. 3–21. Springer, Cham (2023). https://doi.org/10.1007/978-3-031-24159-8_1
10. Łozowicka, D., Kaup, M.: Safety aspects of inland water transport in Poland. TTS **12**, 2016–2020 (2015)
11. Skupień, E., Tubis, A.: The use of linguistic variables and the FMEA analysis in risk assessment in inland navigation. TransNav Int. J. Marine Nav. Saf. Sea Transp. **12**(1), 143–148 (2018)
12. Tubis, A., Skupień, A., Rydlewski, M.: Water transport risk - comparative analysis. In: 8th Carpathian Logistics Congress, CLC2018: Conference Proceedings, 3rd–5th December 2018, Prague, Tanger, Czech Republic, pp. 879–884 (2019)
13. Dong, S.K., Wan, C.Y., An accident causation model for the railway industry: application of the model to 80 rail accident investigation reports from the UK. Saf. Sci. **60**, 57–68 (2013). https://doi.org/10.1016/j.aej.2020.02.030
14. Reason, J.: Human error: models and management. Bmj Clin. Res. **320**, 768–770 (2000)
15. Shappell, S.A., Wiegmann, D.A.: Applying reason: the human factors analysis and classification system (HFACS). Hum. Factors Aerosp. Saf. **1**, 59–86 (2001)
16. Zhan, Q., Zheng, W., Zhao, B.: A hybrid human and organizational analysis method for railway accidents based on HFACS-Railway Accidents (HFACS-RAs). Saf. Sci. **91**, 232–250 (2017). https://doi.org/10.1016/J.SSCI.2016.08.017
17. Yoon, Y.S., Ham, D.H., Yoon, W.C.: A new approach to analysing human-related accidents by combined use of HFACS and activity theory-based method. Cogn. Tech. Work **19**, 759–783 (2017). https://doi.org/10.1007/s10111-017-0433-3
18. Wrocław standards for shaping pedestrian-friendly urban spaces. https://bip.um.wroc.pl/artykul/127/26963/konsultacje-spoleczne-wroclaw-przyjazny-pieszym. Accessed 5 May 2023
19. Wrocław standards of accessibility of urban spaces. https://bip.um.wroc.pl/artykul/1054/62686/wroclawskie-standardy-dostepnosci-przestrzeni-miejskich. Accessed 5 May 2023
20. Resolution of the Wrocław City Council: Cultural Park "Old Town"; Uchwała nr LVI/1465/14 Rady Miejskiej Wrocławia (Dz. Urz. Woj. Doln. z 2014 r. poz. 2049 ze zm.)
21. Polish Standard PN-EN 14206:2006 Statki żeglugi śródlądowej - Trapy dla statków pasażerskich – Wymagania, próby (2006)
22. Polish Standard PN-EN 711:2016-08 Statki żeglugi śródlądowej - Barierki pokładowe – Wymagania, projekty i typy (2016)
23. Polish Standard PN-EN 14504:2019-05 Statki żeglugi śródlądowej - Przystanie oraz pomosty pływające na wodach śródlądowych - Wymagania, próby (2019)
24. Polish Standard PN-EN 14144:2006 Koła ratunkowe - Wymagania, badania (2006)
25. Polish Standard PN-EN 14145:2006 Uchwyty do kół ratunkowych (2006)

26. Polish Standard PN-EN 13573:2004 Statki żeglugi śródlądowej - Systemy kotwiczenia, sczepiania, holowania, manewrowania i cumowania (2004)
27. Polish Standard PN-EN 14503:2006 Statki żeglugi śródlądowej - Porty żeglugi śródlądowej (2006)

Key Factors Influencing the Decision to Buy an Electric Vehicle in Emerging Markets: A Comparison Between University Students' Behavior in Italy and Argentina

Francesca Bruno(✉), Stefano de Luca, Roberta Di Pace, and Facundo Storani

Sustainable Transportation Systems Engineering and Mobility Laboratory (ISTMoS), Department of Civil Engineering, University of Salerno, Fisciano, SA, Italy
{frbruno,sdeluca,rdipace,fstorani}@unisa.it

Abstract. The transition from Internal Combustion Engine Vehicles (ICEVs) to electric vehicles (EVs) is a tricky issue, but particularly challenging in those markets that are characterized by high $ICEV_S$ ownership rates and that may be identified as emerging markets due to the small diffusion of the technology and of the related infrastructures. This study delves into modeling the willingness to buy an EV by young adults within the emerging Italian market, and to draw a comparison with the findings of a previous study carried out on the Argentinian market. To this aim a stated preferences survey was carried out at the University of Salerno (Italy), and then compared with the same survey designed and carried out at the University of Cordoba (Argentina). The Random Utility Paradigm was employed to model the preferences of the respondents, by focusing on the influence of instrumental attributes and psycho-attitudinal factors through a Mixed Binomial Logit formulation and Hybrid Choice model with Latent variables, respectively. The findings suggest that attitudes and perceptions play a substantial and similar role in emerging markets, with only slight differences across the two markets.

Keywords: University students · electric vehicles · emerging markets · latent factors

1 Introduction

Transportation is the fastest growing contributor to global greenhouse gas emissions. Transportation-related CO2 emissions worldwide have increased significantly over the past 50 years, rising from just 2.84 GtCO2 in 1970 to 7.64 GtCO2 in 2021 [1]. This sector is responsible for 20% of global CO_2 emissions and over 75% of them are from road transportation, including cars, trucks, and motorbikes, with passenger cars accounting for more than 40% of road transportation emissions worldwide.

As the world moves towards decarbonization to fight climate change, reducing the use of fossil fuels in homes, businesses, and transportation will be an essential step to achieve energy transition. Electric mobility, combined with renewable energy generation,

G. Sierpiński et al. (Eds.): TSTP 2023, LNNS 907, pp. 46–64, 2024.
https://doi.org/10.1007/978-3-031-53181-1_4

can significantly lower global CO_2 emissions [2]. This strategy not only helps us achieve CO_2 emission targets, but it will also have a positive effect on our health and comfort due to reduced pollution and noise. Hence, electrification, or the process of replacing fossil fuel technologies (such as coal, oil, and natural gas) with electricity, will be a key contributor to sustainability and is set to reshape the global automotive market over the next decade.

In 2021, global sales of EVs reached a new record of 6.6 million, doubling from the previous year. There are now almost eight million EVs on the road worldwide, and this trend will continue to increase. However, EVs have encountered varying levels of success in different regions, including Latin America and Europe.

In this context, there exist several emerging markets, namely markets with small EVs' diffusion rates, not well developed charging infrastructures, and high ICEVs' ownership rates, other than a low perception of EVs' advantages and disadvantages.

These markets are tricky, and it is important to investigate the determinants that affect the prospective purchasing behavior of EVs, as well as understand user perceptions and attitudes towards EVs. To this aim a specific research project was financed by the University of Salerno. The project aimed to investigate and model the purchasing behavior of young adults in emerging markets and involved a dedicated survey in two different emerging markets.

After a first review analysis, the European and Latin America markets were first identified, then the attention was focused on the Italian and Argentinian markets. Indeed, although European EV sales are on par with the global average, there are significant differences across countries. In particular, Italy had a battery electric car adoption rate of 4.6% in 2021 (ISPRA, 2022), significantly lower than leading European countries such as Norway (86%), Iceland (72%), Sweden (43%), the Netherlands (30%), and France (19%). In Latin America, EV sales grew by an average of 75% in 2021, slightly behind the global average of 98%, but the EVs diffusion is small and in Argentina it accounted for only 1.47% in 2021, with the majority (89%) being hybrid electric vehicles (HEVs).

Based on the previous considerations, a stated preferences survey was specifically designed. The first survey was carried out in Argentina [3] on a sample of young potential users (university students) living in the Argentinian emerging market, and valuable insights were drawn both in terms of instrumental and attitudinal factors Then,, also to validate the Argentinian results, the same survey was carried out on a sample of young potential users (university students) living in Italy.

The present paper proposes the results obtained from the Italian survey and compares them with the results gathered from the Argentinian survey. In particular, different random utility models were specified, and socio-psychological factors were explicitly taken into account through the specification of hybrid choice models with latent variables. The role of instrumental attributes was compared, as well as the significance of different latent constructs. The main elements of originality consist in (i) the findings on the Italian context, (ii) in the comparison between the two contexts, and (iii) in the understanding of the role of not directly measurable factors as already observed in several, but different, choice contexts [4–7].

The paper's organization is as follows: Sect. 2 introduces the proposed methodological approach, including the analyzed attributes and the presentation of some preliminary analyses, followed by estimation results in Sect. 3. Section 4 discusses the findings in comparison to the Argentinian context. Finally, Sect. 5 outlines the conclusions, the political entailments and the future perspectives. Appendix provides a brief overview of the survey.

2 Methodological Framework

2.1 Modelling Approach

In this study, the willingness to purchase an Electric Vehicle (EV) has been explored by specifying and calibrating a set of discrete choice models, grounded on random utility theory [8]. In particular, the choice was modeled as a binomial problem where the alternatives were: buying an electric car (alternative1: BUY) and not buying an electric car (alternative2: NOT BUY), therefore choosing the conventional vehicle. Two different types of models were specified: a random parameters mixed binomial logit model (MBNL) and a Hybrid Choice Model (HCM) with latent variables.

In the MBNL model, the systematic utilities [9] were expressed as a linear combination in the attributes, X_{kj}, by means of generic coefficients β_k. The random coefficients, if any, were considered normally distributed with parameters b_k and W_k to be estimated.

$$U_j = V_j + \varepsilon_j = \sum_k \beta_k X_{kj} + \varepsilon_j = \left[\sum_k b_k X_{kj} + W_k X_{kj}\right] + \varepsilon_j \tag{1}$$

Hybrid choice models with latent variables (HCM) allow to take into account not directly measurable factors [10–13]. In particular, latent variables represent latent factors and are defined through indicators collected from attitudinal surveys. Two additional equations should be specified: the structural and measurement equations.

Structural equations represent the latent variables as:

$$LV_p = \gamma_p + \sum\nolimits_l \beta_{SE,l} \cdot X_{SE,lp} + \omega_p \tag{2}$$

where, LV_p is the generic p-th latent variable. It can be specified as a linear combination of users' characteristics attributes, $X_{SE,lp}$, by means of the generic coefficient, $\beta_{SE,l}$, to be estimated, plus an intercept term, γ_p, and a normally distributed error term, ω_p, with zero mean and $\sigma_{\omega p}$ standard deviation.

Measurement equations represent the generic perception indicator $I_{p,n}$. It can be specified as a linear combination of an intercept, $\alpha_{p,n}$, of a latent variable LV_p (by means of a coefficient $\lambda_{p,n}$ to be estimated), and of an error term $v_{p,n}$ which is assumed normally distributed with a mean zero and a standard deviation $\sigma v_{p,n}$:

$$I_{p,n} = \alpha_{p,n} + \lambda_{p,n} \cdot LV_p + v_{p,n} \tag{3}$$

Indicators allow for the identification of latent variables from an econometric point of view. When the indicators are coded as ordered, according to the use of a Likert scale, as in this study, the measurement equation is expressed using an ordered discrete variable I, which assumes values $j_1, j_2, \ldots, j_M$ as follows:

$$I = \begin{cases} j_1 \text{ if } I_{p,n} < \tau_1 \\ j_2 \text{ if } \tau_1 \leq I_{p,n} < \tau_2 \\ \vdots \\ j_i \text{ if } \tau_{i-1} \leq I_{p,n} < \tau_i \\ \vdots \\ j_M \text{ if } \tau_{M-1} \leq I_{p,n} \end{cases} \quad (4)$$

where $\tau_1, \ldots, \tau_{M-1}$ are the parameters to be estimated, satisfying the condition $\tau_1 \leq \tau_2 \leq \cdots \leq \tau_i \leq \cdots \leq \tau_{(M-1)}$

In the HCM, each latent variable (LV) contributes to the systematic utility function:

$$U_j = V_j + \varepsilon_j = \left[\sum_k \beta_k X_{kj} + \sum_p \beta_p LV_{pj} \right] + \varepsilon_j \quad (5)$$

where LV_{pj} represents the unobservable latent variables related to the alternative j, while β_p is the generic coefficients associated with the p_{th} latent variables, to be estimated.

Eight categories of attributes were considered to specify the utility functions. Four of them (also used for the specification of structural equations) were related to directly measurable attributes (Socio-economic characteristics, usual trip type, car engine of the owned vehicle, difference in monthly cost between EV and ICEV), while the remaining 4 relate to the investigated attitudes and perceptions.

Table 1 provides an overview of the attributes belonging to the first 4 subcategories found to be significant in the estimated models; Information related to subcategories 5–8 will be discussed ad detailed in Table 2.

Table 1. Description of significative attributes.

Sub category	Attribute	Description	ITA	
			MIN	MAX
[1]	SEmale	gender (0 = female; 1 = male)	0	1
	SEage	age	19	26
	SEdelta_age	Age (0 = 19 years; 1 = 20 years; 2 = 21 years; 3 = 22 years; 4 = 23 years; 5 = 24 years; 6 = 25 years; 7 = 26 years)	0	7
	SEn_car	Number of vehicles in their household	1	5

(continued)

Table 1. *(continued)*

Sub category	Attribute	Description	ITA	
			MIN	MAX
[2]	SEuse_urb	Main trip type undertaken by the vehicle owned (1 if Urban displacements)	0	1
[3]	SE_alim_diesel	Fuel type of the main vehicle owned (1 if fuelled by Diesel)	0	1
	SEalim_lpg	Fuel type of the main vehicle owned (1 if fuelled by Liquefied Petroleum Gas)	0	1
[4]	VARmonthly_cost	Variation in monthly cost [EUR] between an electric car and a conventional one	0	80

2.2 Preliminary Analysis

Around 400 students from university of Salerno took part in the study, and each participant encountered a varying number (ranging from three to five) of scenarios presented in a randomized order. The survey yielded 292 valid responses, resulting in a total of 1460 observations.

Preliminary descriptive analyses were conducted, and the reliability and validity of four psycho-attitudinal factors were examined through a confirmatory factor analysis.

In terms of gender distribution, males comprise approximately 62.3%, a percentage that closely match the gender distribution of driving licenses the Campania Region (58.8% men) in 2017. In terms of family composition, a majority of respondents (77%) reported having 4 to 5 members. The survey also revealed that a high percentage of students' families owned 2 or more vehicles (82.4%). A notable 64.2% of the respondents possessed diesel-fueled vehicles, a figure almost three times greater than those who owned petrol-fueled vehicles. Additionally, the predominant travel pattern reported by the majority of respondents pertained to suburban trips.

In the survey, 4 non-directly measurable factors were investigated: attitude towards the environment, attitude towards technical features, perception of EVs' advantages, and perception of EVs' drawbacks. Each factor was examined using indirect statements, with participants being asked to express their level of agreement or disagreement on a 5-level Likert scale. For further information and specifics, please refer to the Appendix.

The implementation of latent variable models requires the prior establishment of valid and reliable constructs that accurately represent unobservable attitudes or perceptions. Reliability refers to the consistency of a measure, that is, whether the results can be replicated under the same conditions, while validity refers to the accuracy of a measure, that is, whether the results truly represent what they are supposed to measure.

To this aim, a confirmatory factor analysis (CFA) was conducted to verify the reliability and validity (convergent and discriminant) of the constructs and thus to check whether the information collected from the survey accurately reflects the hypothesized constructs. The internal consistency reliability of the constructs was assessed using Cronbach's alpha [14] and Composite reliability[1] [15, 16].

Convergent validity was assessed referring to the outer loadings of the indicators and to the average variance extracted (AVE) [17]. A satisfactory level of convergent validity is achieved when all item-to-factor loadings are significant, and the AVE score is higher than 0.50 within each dimension. However, an AVE score higher than 0.40 is still acceptable if the composite reliability (CR) is higher than 0.60 [18, 19].

Discriminant validity represents the extent to which a construct differs statistically from other constructs. This study employed three measures of discriminant validity: cross-loading [20], the Fornell-Larcker criterion [18], and the heterotrait-monotrait ratio (HTMT) of correlations [21].

Out of the four investigated constructs, only two successfully met the criteria: the attitude towards technical features and the perception of EVs' advantages. Table 2 presents the outcomes of the confirmatory analysis, exclusively displaying the indicators who saturate the two confirmed constructs (and excluding those eliminated during this analysis). To offer a thorough understanding of the constructs, the table also includes supplementary preliminary statistics such as the mean and standard deviation.

Table 2. Reliable and valid attitudes and perceptions in Italian context

LV	Indicators	mean	sd	OL	AVE	C alpha	CR
LV1: Attitudes toward technical features	TF_{tech_fea}	3.7	0.8	0.71	0.58	0.77	0.84
	TF_{top_speed}	2.9	0.8	0.72			
	TF_{accel}	3.1	0.9	0.81			
	TF_{power}	3.5	0.8	0.78			
LV4: Perception of advantages of EVs	AD_{Red_CO2}	4.1	0.8	0.78	0.41	0.54	0.72
	$AD_{Efficiency}$	3.4	1.1	0.73			
	$AD_{Red\_acustic\_poll}$	3.6	0.9	0.80			
	AD_{Less_parts}	3.0	1.1	0.57			

(continued)

[1] Cronbach's alpha is a conservative measure of reliability. In contrast, composite reliability tends to overestimate the internal consistency reliability. Therefore, true reliability usually lies between Cronbach's alpha (representing the lower bound) and composite reliability (representing the upper bound) (Hair et al., 2017).

Table 2. (*continued*)

Legend	
TF_{tech_fea}	A vehicle's technical features significantly influence my choice in purchasing a new car
TF_{top_speed}	When I am choosing a car, I find myself spending a lot of time checking out differences in terms of top speed (km/h)
TF_{accel}	Immediate acceleration increases my driving pleasure
TF_{power}	I prefer driving a car with a powerful engine
AD_{Red_CO2}	I am interested in EVs so that I can contribute to a reduction in CO2 emissions
$AD_{Efficiency}$	Compared to a normal car, EVs are superior in terms of energy efficiency
$AD_{Red\_acustic\_poll}$	I believe that using EVs can significantly reduce noise pollution in cities
AD_{Less_parts}	I prefer having a car with fewer mechanical components

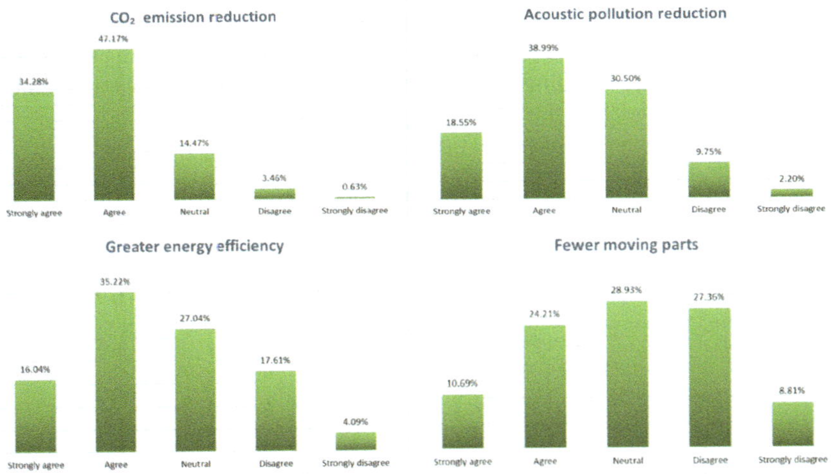

Fig. 1. Level of agreement with the indicators related to the latent variable "Perception of Advantages of EVs"

However, as better discussed later, only one of the two constructs (the perception of EV advantages) was found to be significant for choice modelling. Hence, the distribution depicting the degree of user agreement with statements that saturate this construct is also provided (Fig. 1). In terms of CO_2 emission reduction, more than 81% of respondents express a positive stance, with a majority "strongly agreeing" or "agreeing" with the statement highlighting its significance. Additionally, when it comes to the importance of reducing acoustic pollution, respondents also demonstrate a favorable perspective, with 39% in agreement and 19% "strongly agreeing," although a notable 31% adopt a neutral position. A similar trend emerges with the indicator related to energy efficiency. However, there is a distinct pattern regarding the importance of having a lower number of moving parts, where 29% of respondents maintain a neutral stance, while over 36% express disagreement or do not agree at all.

3 Estimation Results and Comparison with the Argentinian Case Study

This section discusses the results of the finding obtained in the Italian context, considering the specification with two functional forms: a mixed binomial logit model (MBNL) and a Hybrid Choice Model (HCM).

3.1 Mixed Binomial Logit Model Results

The results of the best mixed binomial logit model obtained are presented in Table 3. The same tables also incorporate the results obtained in the Argentinian context. These findings and the comparison will be thoroughly discussed in the upcoming chapter.

As evident from the results, within the Italian context, the decision to purchase or not an electric vehicle is notably influenced by four key attributes: gender, age, variation in monthly ownership costs linked to electric vehicles as opposed to conventional ones, and the type of power source of the respondent's current vehicle.

In particular being male increases the probability of not buying an EV, indicating a greater preference for traditional, performance-oriented vehicles among male students.

Older students seem do not appreciate the possibility to buy EV, indeed, the utility associated with the choice "not buy an EV" increases as the age of the users rises.

The variance in monthly cost decreases the likelihood of buying an EV, as expected. It is the only attribute that exhibits random distribution among users, and it comes with a noteworthy standard deviation of 19%. This emphasizes the existence of distinct preferences among users and varying sensitivity to cost factors, which aligns with previous literature studies that highlight cost as a significant obstacle to EVs' adoption.

As already mentioned, most users are owners of diesel-powered vehicles. This aspect should not be overlooked, as the results of this model show that owning a diesel car also influences the choice, indicating a tendency to stick with ICEVs among users who own a diesel vehicle. This highlights a resistance to transitioning to alternative-fuelled vehicles, indicating a certain reluctance to change, probably due to uncertainties about how new engine technology might affect daily mobility.

Table 3. Results – Mixed Binomial Logit Models (t-test in brackets)

Mixed Binomial Logit Models				
Attribute	ITA		ARG	
	BUY	NOT BUY	BUY	NOT BUY
ASC	+ 5.41		+ 4.120	
	(+9.02)		*(+8.98)*	
SE_age_less25[a]				+ 0.245
				(+1.94)
SE_delta_age		+ 0.192		
		(+1.77)		
SE_male		+ 0.806		+ 0.684
		(+2.71)		*(+3.06)*
SE_n_car[b]				+ 0.146
				(+1.37)
SE_n_family[c]				+ 0.089
				(+2.12)
SE_alim_diesel		+ 0.619		
		(+2.26)		
VAR_monthly_cost		+ 0.133		+ 0.0875
		(+14.51)		*(+15.53)*
S_VAR_monthly_cost		+ 0.025		+ 0.011
		(+3.93)		*(+11.52)*
Models' statistics				
	ITA		**ARG**	
rho-square:	0.614		0.435	
adjusted rho-square:	0.606		0.429	

[a]Age class of individuals (1 = if Age $\leq$ 25; 0 otherwise).
[b]Number of vehicles in their household (In the Argentinian context, minimum value = 1; maximum value = 5).
[c]Number of household members (In the Argentinian context, minimum value = 1; maximum value = 8).

3.2 Hybrid Binomial Logit Model Results

Hybrid choice models were then calibrated by incorporating the valid and reliable latent constructs into the utility function of the alternative "buy an EV". The results of the choice component of the best hybrid binomial logit model obtained are presented in Table 4.

Estimation results indicate that gender, age and variation in monthly cost attributes retain their significance and interpretation from the MBNL model. Additionally, the perception of EVs' advantages (LV4) has a positive effect on the willingness to purchase an EV. This perception is defined by different socioeconomic attributes that can vary the sensibility of the interviewee. Table 5 summarizes the results of the structural model.

As can be seen owning multiple cars in the household increases the perception of EVs benefits as it offers flexibility in choosing vehicles based on specific commuting needs. Urban use also contributes to a greater perception of EVs advantages, as shorter travel distances in urban areas reduce range anxiety and increase confidence in this solution.

Table 4. Estimation results – Utility function of Hybrid Binomial Logit Models (t-test in brackets)

Hybrid Binomial Logit Models				
Attribute	ITA		ARG	
	BUY	NOT BUY	BUY	NOT BUY
ASC	+6.15		2.61	
	(+3.75)		*(+8.09)*	
SE_age		+0.162		
		(+2.23)		
SE_male		+0.698		
		(+3.54)		
VAR_monthly_cost		+0.116		+0.0972
		(+17.7)		*(+20.09)*
B_LV_2[a]			0.434	
			(+1.19)	
B_LV_4[b]	+0.988		0.625	
	(+3.00)		*(+2.64)*	
DELTA TAU_1	+0.521		0.52	
	(+27.3)		*(+39.1)*	
DELTA TAU_2	+1.30		1.25	
	(+31.8)		*(+40.9)*	
Models' statistics				
	ITA		**ARG**	
rho-square:	0.622		0.446	
adjusted rho-square:	0.621		0.443	

[a]LV_2: Attitudes to-wards the environment.
[b]LV_4: Perception of advantages of EVs.

Although the fuel supply of the main owned car loses its significance in the choice function, it plays a role in defining the perception of the advantages of EVs. Indeed, "owning a liquefied petroleum gas (LPG) vehicle" contributes to an increased perception of the benefits of EVs, suggesting that users with conventional vehicles are more resistant to switching to EVs compared to those with less conventional fuel vehicles, like LPG.

Table 5. Estimation results – Structural model of HCMs

ITA	Structural Model - LV4: advantages of EVs				
	β MEAN4	ω 4	SE_alim_lpg	SE_use_urb	SE_n_car
	+ 1.31 (+21.5)	+ 0.41 (+7.01)	+ 0.099 (+1.84)	+ 0.054 (+1.84)	+ 0.115 (+1.67)
ARG	**Structural Model – LV2: Attitudes towards the environment**				
	β MEAN2	ω 2	SE_male	SE_age_30_34[a]	SE_comp_car[b]
	+ 0.431 (+6.75)	+ 0.692 (+16.8)	-0.663 (-10.4)	+ 0.459 (+1.62)	-0.24 (-3.9)
	Structural Model - LV4: advantages of EVs				
	β MEAN4	ω 4	SE_male	SE_age_under25	SE_alim_gnc[c]
	+ 1.56 (+15.5)	+ 1.25 (+23.1)	-0.604 (-8.14)	-0.163 (-2.03)	+ 0.146 (+1.48)

[a] Age class of individuals. (1 = if 30 ≤ Age ≤ 34; 0 otherwise).
[b] The habit of comparing two or more cars when deciding to buy (1 = yes; 0 = no).
[c] Fuel type of the main vehicle owned by the interviewee (1 = if fuelled by compressed natural gas).

Table 6. HCMs Estimation results – Measurement model of HCM

Measurement model							
ADRed_CO2		ADEfficiency		ADRed_acustic_poll		ADsimplicity	
α 41	0	α 42	−2.79 (−5.65)	α 43	−0.447 (−2.59)	α 44	−3.05 (−6.18)
λ 41	1	λ 42	2.36 (6.59)	λ 43	0.864 (7.51)	λ 44	2.17 (6.01)
V41	1	V42	0.926 (15.5)	V43	1.14 (28.3)	V44	1.1 (18.6)

As regards the measurement model, for identification purpose, the parameters α, λ, and v of one indicator are normalized to zero and one. All signs of the statements (presented in Table 2) are consistent with interpretative expectations.

4 Italy vs Argentina: Discussion and Comparison

From preliminary analysis, it results that the two samples exhibited similarities. First of all, the Argentinian sample is composed by a majority of males (65.2%) that closely match the gender distribution of driving licenses in Buenos Aires (67.4% men) in 2017. In both the examined context the respondents were primarily young adults, the target of this analysis. In terms of family composition, a majority of respondents in both countries reported having 4 to 5 members (61% in Argentina and 77% in Italy). The survey also revealed that a high percentage of Argentinean students' families owned 2 or more vehicles (61.67%), while in Italy, this rate was even higher (82.4%).

The most significant differences between the two samples are related to the power supply of the owned vehicles and the usual type of trips undertaken. The vehicle fleet composition differs, with gasoline-fueled vehicles being dominant in Argentina, and diesel-fueled vehicles being more prevalent in Italy. Regarding trip types, urban trips were more common in Argentina, while suburban trips were more frequent in Italy. It is worth noting that due to longer distances between cities in Argentina, journeys of 100 km or more are considered "extra urban," whereas in Italy, distances as short as 10 km fall into this category.

Finally, in contrast to the Italian context, within the Argentinean context, all four of the latent factors under investigation demonstrated reliability and validity. This enables the consideration of the influence of multiple latent variables in shaping users' choices.

In both case studies, the MBNL (Table 3) results reveal that gender, age, and variation in monthly cost are statistically significant. Specifically, being male increases the probability of not buying an EV, indicating a greater preference for traditional, performance-oriented vehicles among male students, independently from the geographical context.

The role of age differs in the two contexts, with younger students in Argentina appearing less interested in purchasing more sustainable vehicles, while older students in Italy show lower utility associated with EV purchases.

Regardless of context, the variance in monthly cost decreases the likelihood of buying an EV, as expected. The standard deviation of monthly cost is similar in both case studies, with a deviation of 13% in Argentina and 19% in Italy.

The two models differ for some attributes. The number of family members and the number of cars owned in the household are significant only in the Argentinean survey, enhancing the utility linked to choosing not to purchase an EV. This suggests that larger households and those with more cars exhibit a preference for internal combustion engine vehicles (ICEVs), probably due to perceived affordability and risk aversion.

In the Italian context, also the power of the owned vehicle influences the choice, and highlights an inertia to the transition towards alternative-fuelled vehicles.

Results from the HCMs (Table 4) show that in the Argentina context, some of the sociodemographic attributes that were statistically significant in the MBNL model lose their significance in the choice model but regain significance in the structural one. A positive impact on the inclination to buy an EV is observed when there is a favorable attitude toward the environment and a positive perception of the advantages offered by EVs. On the other hand, In Italy, gender and age attributes retain their significance and interpretation observed in the MBNL model, but only the latent variable related to the perception of EVs' advantages plays a significant role in the choice process.

Looking at the results of the structural model (Table 5), in the Argentinean context, the male gender has a negative influence on both latent variables, aligning with previous literature that suggests a preference for conventional vehicles among males due to perceived performance advantages. Age and the habit of comparing technical features play a role in shaping the underlying environmental attitude. In the case of the examined sample, an older age exhibits a positive effect on environmental attitudes, suggesting that a more mature population tends to exhibit greater awareness of environmental concerns. On the other hand, users who prioritize technical characteristics over environmental concerns have a weaker attitude towards the environment, preferring high-performance vehicles.

Remaining in the Argentinian context, in addition to gender and age, the fuel supply of the main owned car plays a role in the latent variable "perception of the advantages".

This attribute proves significant both in the Italian and in the Argentinian context, with a similar role associated with the possession of a less conventional fuel vehicle, represented by compressed natural gas (CNG) in the Argentinian context and by liquefied petroleum gas (LPG) vehicle in Italy.

In Italy, other attributes (the number of car and the main use of the car) are significant in shaping the perception of advantages of EVs.

5 Conclusions

Environmental impact concerns have become a ubiquitous issue, and the ongoing challenge for the transportation sector is to reduce the adverse effects it imposes. Numerous strategies and solutions are being put forth to address this challenge, encompassing initiatives on both the supply [22, 23] and demand [24, 25] sides of transportation. Within this context, the integration of EVs into the market stands out as one of the proposed supply-side solutions. However, it is equally vital to delineate policies designed to encourage the adoption of EVs, recognizing that these policies may need to be tailored to the specific circumstances of the region or area under examination.

The decision to purchase an EV cannot be solely explained in terms of instrumental and socio-economic attributes, as attitudes and perceptions also play a significant role.

Technologically, the high purchase price of EVs compared to internal combustion vehicles is a major deterrent for users. Heterogeneity among respondents, as indicated by the MBNL models, is explained by the monthly cost, highlighting the impact of perceived monetary cost on behavior.

Several characteristics closely connected to users also play a significant role in the purchasing decision. It is worth noting that, while two different case studies have shown the significance of certain attributes and latent variables, their relative importance depends on the socio-cultural context, market maturity and awareness of EVs technology. According to the results we may further classify immature markets in "aware" and "not-aware" indicating different levels of technology perception due to experience.

Specifically, "aware" users possess a well-defined perception of EVs, but electric mobility has minimal market presence and lacks sufficient technological infrastructure. Conversely, "not-aware" users are familiar with the technology, but EVs are not yet really available in the market, and there is a lack of charging infrastructure. In this study, Italy represents an aware market, while Argentina represents a not-aware market.

An important consideration lies in the fact that, in both contexts, the perception of the advantages of electric vehicles plays a key role in the purchase decision. Strategies and policies should aim to enhance this perception, emphasizing the benefits of EVs such as reduced emissions, noise pollution, higher energy efficiency, and lower maintenance costs (due to greater simplicity of the involved mechanical parts). Such perception is linked to socioeconomic characteristics and user-related circumstances. However, it should be noted that in a "not aware" context such as the one of Argentina, this latent variable is not the only factor contributing to the purchase decision, but also the respondents' personal attitude towards the environment proves to be significant. This point is crucial because it suggests that in aware market there is less "randomness" associated with personal attitudes in the definition of utility associated with the electric purchase alternative; hence, it emphasizes, the greater effectiveness that the intervention strategies might encounter in a more aware market, rather than in a less aware one. At the same time, it also emphasizes, the role that natural attitudes of "innovators" have in influencing the triggering of a new market. This point is fully consistent with the diffusion of innovation theory proposed by Rogers [26], that defines diffusion as "the process in which an innovation is communicated thorough certain channels over time among the members of a social system" and with the Bass model [27], according to which the diffusion of a good is closely linked to the innovation coefficient (which represents external influences and advertising) and to the imitation coefficient (which represents internal influences that lead to adoption by imitation and word of mouth). Actually, in line with the categorization of adopters proposed by the Diffusion of Innovation Theory [26], the findings suggest that in a less conscious market such as the one of Argentina, it is still the so-called 'innovators' who choose an electric vehicle, while in the Italian context, the choice has also extended to the following categories, the so called "early adopters".

Another point of interest is that certain socio-economic characteristics directly or indirectly influence the decision to purchase an EV. Specifically, it emerges that in the Argentinean contest, gender and age do not directly influence the purchase choice, but rather the attitude towards the environment and the perception of EV benefits. In Italy, these characteristics have a direct effect on the purchase choice. Consistent with other studies [28, 29], it emerges that males and age pose obstacles to the purchase of electric vehicles. Typically, males tend to prefer traditional cars that are associated with higher performance (in terms of engine power and top speed). The age effect is frequently reported in the literature, but unlike gender, there is no specific trend. Some studies show a greater propensity towards electric vehicles for younger users [30] while other studies report that the tendency towards electric vehicles is characteristic of middle-aged people [31].

These findings provide a general market segmentation overview, guiding future research and marketing strategies. To gain a more comprehensive understanding, further studies targeting different segments would help to determine the most effective marketing strategies.

Although this study offers valuable insights it is important to acknowledge its limitations. The research is centered around a specific cohort of young adults (university students), potentially constraining the extent to which the findings can be applied more broadly. However, it is worth noting that research has shown that potential EV buyers tend to be more affluent and educated [32], indicating that this population subset is relevant for exploring this topic.

Appendix: Experimental Set-Up - Survey Description

This paper presents the findings of a study conducted Argentina and Italy, investigating the factors influencing the adoption of electric vehicles (EVs).

In light of the ongoing growth of the electric vehicle market in Italy and Argentina, a forward-looking study based on data from a revealed preference survey appeared premature. As a result, the willingness to purchase an EV was investigated through a stated preference (SP) experiment. The survey was conducted online in 2019 through a web-based questionnaire that took no more than fifteen minutes to complete and that was administered to students at University of Cordoba in Argentina and to the University of Salerno in Italy.

The study employed a specially designed SP questionnaire consisting of five sections and included an initial filter section to assess participants' intention to purchase a car within the next 18 months. Those who did not express an interest were directed to the exit section of the survey.

The first two sections of the survey aimed to collect observable variables that could be relevant in characterizing potential future purchasers of electric vehicles, in terms of their socioeconomic backgrounds (gender, age, household size, and number of cars owned), as well as their usual travel habits (fuel type of the owned car, annual kilometers driven, and primary use of the vehicle, i.e., urban or extra-urban). This information could be included in both the specification of the utility function and in the specification of the latent variables.

The third section (Fig. 1) explored the influence of two attitudes (towards technical features and the environment) on respondents' willingness to purchase an electric vehicle EV, using a five-point Likert [33] scale (from strongly disagree to strongly agree).

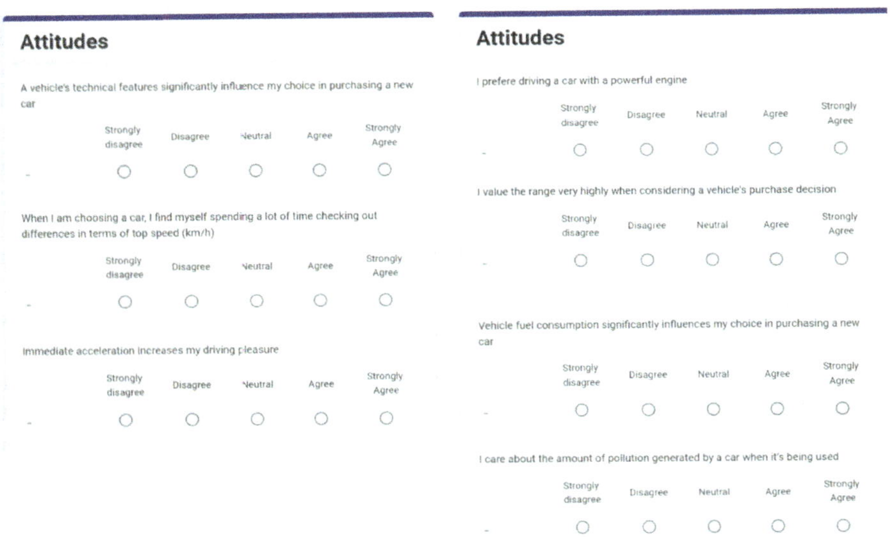

Fig. 2. Third survey section about attitudes toward the environment and technical features

In the fourth section, participants were presented with different choice scenario for purchasing a new vehicle, with a comparison between an electric vehicle (EV) and a conventional internal combustion engine vehicle (ICEV). This section incorporated an educational module aimed at clarifying the basic principles of electric vehicle technology. The aim was to ensure that the possible propensity to purchase an electric vehicle was based on information that was both reliable and equally accessible to all survey participants. The main features discussed were: zero tailpipe emissions, quiet operation, possible charging locations (private or public), charging times (ranging from 30 min to 8 h), range (minimum of 200km for standard cars, enough for a week of urban trips) reduced maintenance costs (due to fewer moving parts), motor information (same power as combustion engines, but lighter). Then, each respondent was faced with choices.

Each respondent was presented with two cars: a Renault Clio (alternative A, ICEV) and a Renault Zoe (alternative B, EV). The cars were from the same manufacturer, similar in size, color and main features to minimize the impact of these characteristics on the choice decision (Fig. 2.a). It was hypothesized that the interviewee had enough budget to purchase either car, and a travel scenario was set (urban and 40 km/day).

The monthly cost of the conventional vehicle was described (see Fig. 2. b), and then respondents were presented with different scenarios with varying monthly costs of the electric vehicle relative to the conventional one. The five scenarios were equal monthly cost, + 10%, + 20%, + 30%, and + 40%.

More details on the survey description are provided in [3].

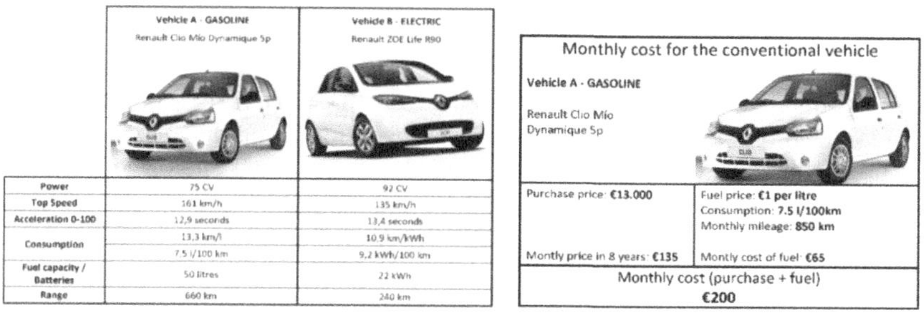

(a) (b)

Fig. 3. Fourth survey section: (a) Comparison between Renault Clio and Renault Zoe; (b) Example of a typical choice scenario

In the fifth section, we gathered insights from users regarding their perception of advantages and disadvantages of electric vehicles[3, 34] which could influence their inclination to buy such vehicles. Similarly to the third section, using a five-point Likert scale, respondents were asked to express their level of agreement or disagreement with some indirect statements (Fig. 3).

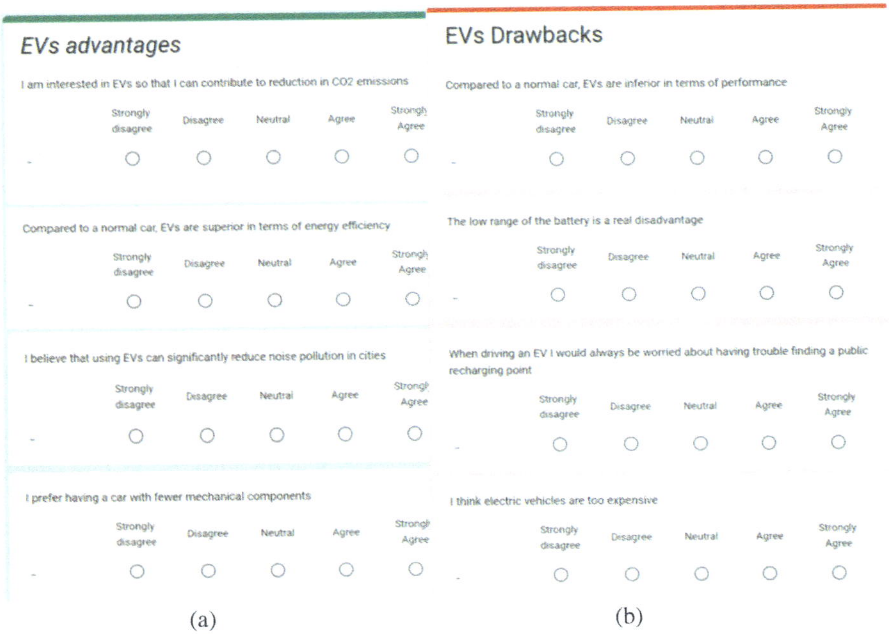

Fig. 4. Fifth survey section: (a) Investigation of perception of EV advantages; (b) Investigation of perception of EV drawbacks.

Around 1,000 participants from two universities took part in the study, and each participant encountered a varying number (ranging from three to five) of scenarios presented in a randomized order. This approach was employed to prevent any serial correlations and ensure the independence of all choices made by participants. The survey yielded 467 valid responses from the University of Cordoba and 292 from the University of Salerno, resulting in a total of 2,148 observations for the Argentinian survey and 1,460 for the Italian survey.

References

1. IEA - Global EV Outlook 2022 – Analysis. https://www.iea.org/reports/global-ev-outlook-2022. Accessed 26 Apr 2023
2. de Luca, S., Di Pace, R.: Aftermarket vehicle hybridization: Potential market penetration and environmental benefits of a hybrid-solar kit. Int. J. Sustain. Transp. **12**, 353–366 (2018). https://doi.org/10.1080/15568318.2017.1377325
3. de Luca, S., Storani, F., Bruno, F., Di Pace, R.: Adoption of electric vehicles by young adults in an emerging market: a case study from Argentina. Transp. Plann. Technol., 1–32 (2023). https://doi.org/10.1080/03081060.2023.2265362
4. de Luca, S., Di Pace, R.: Evaluation of Risk Perception in Route Choice Experiments: An Application of the Cumulative Prospect Theory. IEEE Conference on Intelligent Transportation Systems, Proceedings, ITSC. 2015-October, 309–315 (2015). https://doi.org/10.1109/ITSC.2015.60

5. Bifulco, G.N., Di Pace, R., Viti, F.: Evaluating the effects of information reliability on travellers' route choice. Eur. Transp. Res. Rev. **6**, 61–70 (2014). https://doi.org/10.1007/S12544-013-0110-4/TABLES/8
6. Henke, I., Pagliara, F., Cartenì, A., Coppola, P.: The Impact of COVID-19 Pandemic on Public Transport: A Mobility Survey in Naples (South of Italy). The Open Transportation J. **17**, (2023). https://doi.org/10.2174/18744478-V17-E230420-2022-47
7. Tsouros, I., Polydoropoulou, A.: Who will buy alternative fueled or automated vehicles: a modular, behavioral modeling approach. Transp. Res. Part A Policy Pract. **132**, 214–225 (2020). https://doi.org/10.1016/J.TRA.2019.11.013
8. Domencich, T.A., McFadden, D.: Urban travel demand: a behavioral analysis : a Charles River Associates research study (1975)
9. Train, K.E.: Discrete choice methods with simulation, second edition. Discrete Choice Methods with Simulation, Second Edition. 9780521766555, 1–388 (2009). https://doi.org/10.1017/CBO9780511805271
10. Walker, J.: Extended Discrete Choice Models: Integrated Framework, Flexible Error Structures, and Latent Variables (2001). http://transp-or.epfl.ch/courses/dca2012/WalkerPhD.pdf
11. Ben-Akiva, M., et al.: Extended framework for modeling choice behavior. Mark. Lett. **10**, 187–203 (1999). https://doi.org/10.1023/A:1008046730291
12. Bahamonde-Birke, F.J., Kunert, U., Link, H., Ortúzar, J. de D.: About attitudes and perceptions: finding the proper way to consider latent variables in discrete choice models. Transp. (Amst). **44**, 475–493 (2017). https://doi.org/10.1007/s11116-015-9663-5
13. De Luca, S., Di Pace, R.: Did attitudes interpret and predict "better" choice behaviour towards innovative and greener automotive technologies? a hybrid choice modelling approach. J Adv Transp. 2020 (2020). https://doi.org/10.1155/2020/5135026
14. Cronbach, L.J.: Coefficient alpha and the internal structure of tests. Psychometrika **16**, 297–334 (1951). https://doi.org/10.1007/BF02310555
15. Jöreskog, K.G.: Statistical analysis of sets of congeneric tests. Psychometrika **36**, 109–133 (1971). https://doi.org/10.1007/BF02291393/METRICS
16. Werts, C.E., Linn, R.L., Jöreskog, K.G.: Intraclass Reliability Estimates: Testing Structural Assumptions. https://doi.org/10.1177/001316447403400104. 34, 25–33 (1974). https://doi.org/10.1177/001316447403400104
17. Hair, J.F., Hult, G.T., Ringle, C., Sarstedt, M.: A Primer on Partial Least Squares Structural Equation Modeling (PLS-SEM) - Joseph F. Hair, Jr., G. Tomas M. Hult, Christian Ringle, Marko Sarstedt (2017)
18. Fornell, C., Larcker, D.F.: Evaluating structural equation models with unobservable variables and measurement error. J. Mark. Res. **18**, 39 (1981). https://doi.org/10.2307/3151312
19. Huang, C.-C., Wang, Y.-M., Wu, T.-W., Wang, P.-A.: An empirical analysis of the antecedents and performance consequences of using the moodle platform. Int. J. Inf. Educ. Technol., 217–221 (2013). https://doi.org/10.7763/IJIET.2013.V3.267
20. Chin, W.W.: The partial least squares approach to structural equation modeling the proactive technology project recovery function: a methodological analysis view project research methods view project. Modern Methods Bus. Res. **295**, 295–336 (1998)
21. Henseler, J., Ringle, C.M., Sarstedt, M.: A new criterion for assessing discriminant validity in variance-based structural equation modeling. J. Acad. Mark. Sci. **43**, 115–135 (2015). https://doi.org/10.1007/s11747-014-0403-8
22. Cartenì, A., Henke, I., Mallozzi, F., Molitierno, C.: A multi-criteria analysis as a rational evaluation process for building a new highway in Italy. WIT Trans. Ecol. Environ. **217**, 713–723 (2018). https://doi.org/10.2495/SDP180601

23. de Luca, S., Di Pace, R., Memoli, S., Pariota, L.: Sustainable traffic management in an urban area: An integrated framework for real-time traffic control and route guidance design. Sustainability (Switzerland). 12 (2020). https://doi.org/10.3390/SU12020726
24. de Luca, S., Papola A.: Evaluation of travel demand management policies in the urban area of Naples (2001)
25. Cascetta, E., Carteni, A., Henke, I.: Acceptance and equity in advanced path-related road pricing schemes. In: 5th IEEE International Conference on Models and Technologies for Intelligent Transportation Systems, MT-ITS 2017 – Proceedings, pp. 492–496 (2017). https://doi.org/10.1109/MTITS.2017.8005722
26. Rogers, E.M.: Diffusion of Innovations, 5th Edition (Google eBook). 576 (2003)
27. Bass, F.M.: A New Product Growth for Model Consumer Durables. Manage. Sci. **15**, 215–227 (1969)
28. Ji, D., Gan, H.: Effects of providing total cost of ownership information on below-40 young consumers' intent to purchase an electric vehicle: a case study in China. Energy Policy **165**, 112954 (2022). https://doi.org/10.1016/J.ENPOL.2022.112954
29. Tian, X., Zhang, Q., Chi, Y., Cheng, Y.: Purchase willingness of new energy vehicles: A case study in Jinan City of China. (2021). https://doi.org/10.1016/j.regsus.2020.12.003
30. Sovacool, B.K., Kester, J., Noel, L., de Rubens, G.Z.: The demographics of decarbonizing transport: the influence of gender, education, occupation, age, and household size on electric mobility preferences in the Nordic region. Glob. Environ. Chang. **52**, 86–100 (2018). https://doi.org/10.1016/J.GLOENVCHA.2018.06.008
31. Lee, J.H., Hardman, S.J., Tal, G.: Who is buying electric vehicles in California? Characterising early adopter heterogeneity and forecasting market diffusion. Energy Res Soc Sci. **55** (2019). https://doi.org/10.1016/j.erss.2019.05.011
32. He, X., Zhan, W., Hu, Y.: Consumer purchase intention of electric vehicles in China: the roles of perception and personality. J. Clean. Prod. **204**, 1060–1069 (2018). https://doi.org/10.1016/J.JCLEPRO.2018.08.260
33. Likert, R.: A technique for the measurement of attititudes. Arch. Psychol. PsycINFO. **22**, 5–55 (1932)
34. Adhikari, M., Ghimire, L.P., Kim, Y., Aryal, P., Khadka, S.B.: Identification and Analysis of Barriers against Electric Vehicle Use. Sustainability 2020, 12, 4850. 12, 4850 (2020). https://doi.org/10.3390/SU12124850

Relationship Among Socio-demographic Characteristics, Activity-Travel Participation, Travel Parameter, Physical Activity Intensity, and Health Parameters

Mujahid Ali(✉) and Elżbieta Macioszek

Department of Transport Systems, Traffic Engineering and Logistics, Faculty of Transport and Aviation Engineering, Silesian University of Technology, Krasińskiego 8 Street, Katowice, Poland
mali@polsl.pl

Abstract. Recent and past studies clearly mentioned the influence of health on activities. However, according to the author view and recent studies, there has been limited studies which shows the influence of daily activities on health parameters. Besides, activity participation is complex which don't influence health directly. Therefore, the current study aim to investigate the influence of daily time-use and activity travel participation (ATP) on health parameters using intensity of physical activity at leisure time as intermediate variable using multi-dimensional three-week household survey. 191 houses and 732 persons in all were counted, which corresponds to 0.029% of the BMA's entire population. The association between activity patterns and health factors is validated using the hierarchical Structural Equation Model (SEM). According to the projected results, a minute increase in public transport mode makes a 0.088 positive correlation with physical health and 0.090 with social health, whereas the total travel time of 0.05 negatively affects physical health. Additionally, there is a strong correlation between improved physical and social health and the accessibility and availability of essential facilities within walking distance. This study employed interdisciplinary techniques that are required to be built to achieve better transportation policy and a healthier society with a higher quality of life.

Keywords: activity travel participation · travel parameters · physical intensity · physical health

1 Introduction

Research on transportation geography and travel behavior, which describes activity-travel participation such as physical, social, economic, and geographic factors, is an area of research where theories are crucial [1]. The time-geographical framework is a more comprehensive and potent way to examine how people behave [2]. Hägerstrand developed the time geography theory, also known as the time-space prism theory, in the 1970s to explain the spatio-temporal limitations that influence people's activity and travel

G. Sierpiński et al. (Eds.): TSTP 2023, LNNS 907, pp. 65–81, 2024.
https://doi.org/10.1007/978-3-031-53181-1_5

decisions [3]. The theory is built on three mechanism: resources, needs, and constraints [4]. Resources are the source of supplies that allow us to move and engage in activities [5]. The human demands of needs and desires are what might motivate someone to engage in an activity [6]. Limitations, which are further separated into coupling, authority, and capacity limitations, limit an individual's ability to move around and engage in an activity [7]. These constraints interact with one's daily time-space prism's requirements and available resources. Individuals, however, have distinct daily requirements, goals, and limitations. They also have various time-space routes and prisms from day to day [8].

A person needs a home, good health, physical health, and social security in order to live a full and active life [9]. As significant as the rights to food and shelter is the right to health. To have access to these necessities, one needs financial resources, which allow one to travel and engage in activities [10, 11]. The action produces the travel behavior. Activity-travel patterns must be examined in order to study travel behavior. Research on travel behavior looks at how people migrate physically for whatever reason away from their home base [12]. Everybody engages in a unique set of activities at certain locations on each given day. People must travel in order to take part in a varied set of activities on any given day [13]. Moving from one location to another permanently requires travel [3, 14].

It is crucial to understand that the transportation system seeks to enable movement for both people and products to get to their destinations. The freight movement is only addressed as far as the people transport goods/cargo for shopping, work, or other restricted purposes [15, 16]. The distribution and production system of services and goods are ignored from the logistic of this study. Keeping it to only passenger transportation permits travel and participation in a variety of other activities [17]. To control the restrictions while engaging in an activity, there must be some requirements, preferences, and resources [5]. Infrastructures including commercial areas and transportation networks, as well as access to a variety of modes of transportation, would make it easier to combine many/few activities with a variety of distinct locations within a certain amount of time [1, 18].

Each individual is different from another and their daily activities are varying; therefore, the activities are categorized as mandatory and discretionary activities [19]. Maintenance and leisure activities are the sub-categorization of Discretionary activities. Those activities which are fixed and difficult to be re-scheduled is called mandatory activities [5, 20, 21]. Besides, discretionary activities are ones that have greater time-based and spatial flexibility and are more readily rescheduled [22, 23]. These pursuits are further separated into indoor and outdoor pursuits. A person's time spent engaging in an activity away from home is usually referred to as "out-of-home activity," whilst their time spent engaging in an activity within their house is the reverse [24, 25].

The entire trip time is influenced by the sort of activity, the location's accessibility, personal obligations, and the available transit mode [26, 27]. It is generally accepted that putting the activity location nearer to the home site and promoting non-motorized transportation reduces an individual's journey time. While some of the activities are far away, others are within walking distance. While some travel for optional, leisure, and maintenance reasons, others travel for tasks that are required. These journeys are made for both indoor and outdoor activity. On any given day, people participate in a variety of activities at various locations using a variety of modes of transportation, including non-motorized, public transportation, and motorized. One's social, physical, and emotional health might be positively or severely impacted by how they commute to and from work [28–31].

Health and transportation are connected at different stages. Sometimes decisions about transportation have an impact on health, and vice versa [32]. Consequently, health is a capacity restriction when engaging in an activity that has the potential to affect activity-travel patterns [25]. Health indicators will provide a new aspect of personal capacity variables. An individual wants to travel and take part in an activity at a new location and has requirements, desires, and accessible resources like a car and money. However, owing to social (such as peer pressure from family and friends), physical (such as blindness, broken legs, disorder, and illness), and mental health (such as stress), he or she is incapable to move and engage in an action [25, 33].

The conceptual model depicts the main link that is crucial for the interaction between travel behavior and health, but it does not fully account for all categories of factors, including those that might also have an impact on health parameters, such as multitasking, lifestyle choices, genetics, drinking, smoking, etc. These aspects are not covered in detail in this work due to their intricacy and the need for delineation. Relationships between time-use activity-travel participation, travel habits, and health are all included in the model and are connected to several intermediary variables. Studying how travel behavior (TB) and activity-travel participation (ATP) affect health is a very complex phenomena [22, 34]. According to the author and contemporary research, no study has been done to directly link individuals' time-use and ATP and health conditions [17, 35–38]. In addition, activity-travel patterns are too broad a term to accurately describe the health parameters [39–45]. he relationship between activity-travel patterns and health indicators can be found with the use of more precise variables. The relationship of activity-travel patterns and health indicators is investigated using active transportation and physical intensity as a mediation variable. Health parameters can be improved and maintained by physical activity [46, 47], however the best method, intensity, frequency, and duration of physical exercise have not been identified. Furthermore, other studies have unmistakably demonstrated that health is a constant restriction that affects activity-travel participation and travel characteristics [22]. However, the prior study almost ever examined the impact of engaging in a varied set of activities and travel-related factors on health parameters, especially physical and social health, on a given day [34, 35, 48].

To further understand how activity-travel patterns affect health issues, a comprehensive dataset at the individual and household levels is needed. In addition to the advanced BMA problem, this study emphasizes the (1) time-use and activity diary survey. The ability to analyze the interaction between activity-travel participation and health parameters at the household level will be made possible by a comprehensive dataset, which will also give a broad understanding of the activity-travel decision-making process. This will allow us to understand how these factors affect the implementation of policy for various transport modes.

The paper is structured as follows: Sects. 2 and 3 present the study area, research strategy, and dataset. In part 4, exploratory analysis of multidimensional data is offered, and in Sect. 5, the conclusion and debates are covered.

2 Sampling and Population

2.1 Population

In this study the total citizen, working and living people in a specific city of Indonesia are under the consideration, with an age of fifteen years old and greater. The number of samples is needed to be concerned to avoid the chances of error. To determine how many samples should be used, Daryle W. Morgan and Robert V. Krejcie in 1970s introduced a formula, refer to Eq. 1.

$$s = X^2NP\,(1 - P) \div d^2(N - 1) + X^2P\,(1 - P) \tag{1}$$

s = sample size

X^2 = degree of freedom at the required CI.

N = Size of population.

P = the population proportion (assumed to be 0.50).

d = the degree of accuracy expressed as a proportion (0.50).

d = the degree of accuracy expressed as a proportion (0.50).

By using the above formula to determine the number of sampling to be done from a certain population which will represent the whole population in that specific area.

2.2 Sampling Frame

The Bandung Metropolitan Area (BMA) is the study's sample area, and data were gathered over the course of 21 consecutive days from 191 households within the BMA totaling 740 household members who were fifteen years of age or older. Since the household serves as the survey's basic unit, every member of the household was contacted for an interview. Any new legislation that may significantly differ from the gathered dataset during the period gaps after the dataset has been collected have not been discovered. As a result, the 2013 BMA dataset was still sufficient for our investigation. 545 respondents are examined after the respondents with a large number of missing values are excluded. Table 1 provides a description of the responders.

Due to the low participation of residents in Indonesia and other developing countries, the surveys were conducted in Bahasa Indonesia rather than in English. The survey process started with direct interaction between surveyors and potential respondents through random selection, with community pioneers intervening in selected neighborhoods. The BMA dataset includes multidimensional information on the household, physical activity and way of life, as well as information on built environment, time-use and activity diaries, subjective well-being, and a person's subjective traits. Other research that looked at people's activity-travel behavior from multidisciplinary perspectives have used the data set (e.g. Dharmowijoyo, et al. 2015; Dharmowijoyo, et al. 2014; Susilo & Liu).

Table 1. Description of respondent's socio-demographic characteristic

Socio-demographic Variables	Percentage/mean
Male	53.5%
Student	32.0%
Worker	40.0%
Non-worker	26.5%
Age 15–22 (Years old)	11.50%
Age 23–44 (Years old)	42.50%
Age 45–55 (Years old)	11.00%
Older than 55 years	9.50%
High household income (>6 million)	7.5%
Medium household income(3–6 IDR)	92.5%

It is crucial to understand that the primary goal of the transportation system is to give access to destinations for both the products and transportation of people. Passengers and those people who are above fifteen years of age are considered. Both the random and convenience sampling techniques are used for collecting the data in Bandung Metropolitan Area (BMA).

3 Data

Data is a collection of quantitative or qualitative variable values that play a key part in statistical analysis. Data only becomes information that can be used to make decisions when it takes the shape of facts or figures from which inferences may be made. Exploratory research will be carried out for this study, in which both the qualitative and quantitative data will be collected and will be analyzed. In this research some of the variables from the previous data set namely "The 2013 Bandung Metropolitan Area" (BMA) were used. Data were collected in Bandung Metropolitan Area (BMA), Indonesia in 2013 from face to face questionnaires survey, consist of three different set of survey forms such as individual and household survey, time use and activity diary survey, and household physical activity survey in which different set of questions were asked such as individual and household demographics, physical activities, intensities of physical activities, travel behaviour, lifestyle habits, social activities and health promotion quality of life using short from SF-36.

3.1 Household Dataset

The household data set included socio-demographic variables such as particular information of household members, social and communication activities with or with other household members, perceived measurement of accessibility, neighbourhood information and household questions. Social and communication was consumed for collecting how many days in a week (frequency) and how much hours in a day (intensity) individual commenced social and family interactions. However, due to highly congested traffic in BMA, perceived approachability was used instead of objective approachability measurements.

3.2 Diary Activity Travel Participation

The dataset included diary activity travel participation which were collected by ATP. Each day, there were 96 slices of this survey, spaced 15 min apart. Activity travel participation were consisting of 23 different types of activities which were categorized as in-home and out-of-home mandatory activities, discretionary leisure, and maintenance activities.

3.3 Health Related Quality of Life (QoL)

The survey also included a set of health related QoL indicators and their influencing factors which was based on SF-36 (Short-Form 36) the most widely used generic measurement of health related survey and are adopted by more than 110 countries and consist of 36 health related quality of life indicators and their influencing factors, including healthy lifestyle habits, health activities, activity-travel behaviour and living environment. The SF-36 is contained of eight subscale which measure health parameters such as mental physical, and social health which are physical functioning (PF), limitation on role functioning because of physical health (RP), body pain (BP), general health (GH), vitality (VT), social functioning (SF), limitation on role functioning because of emotional health (RE), and mental health (MH). Physical health were defined from PF, RP, and BP whereas RP, SF, RE defined social health and SF, BP, GH, VT, and MH as a mental health as shown in Fig. 1. For more complete information regarding SF-36 can be found in a paper by [21, 33].

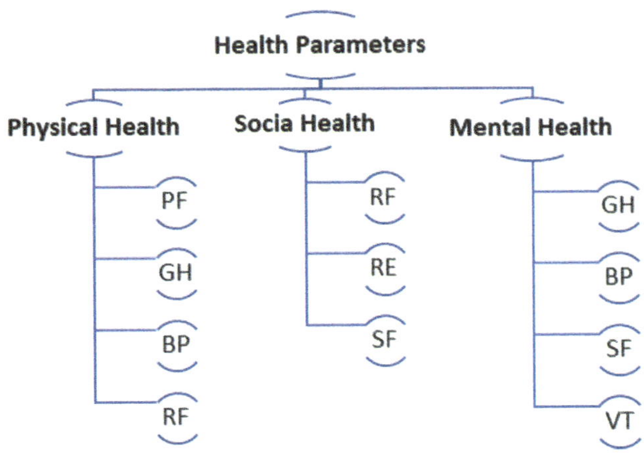

Fig. 1. Health related quality of life with factors

4 Data Analysis

The qualitative data/materials was collected from the participants through questionnaires survey on face to face interview from those who had stated their willingness to be interviewed. The data set was mainly divided into three sections. (1) Household survey which is employed for finding socio-demographic variables such as gender, income and so on. (2) Time-use and activity dairy survey (3) and Household physical activity survey which was consist of and physical social activities of an individual and health related quality of life (Qol). Different analysis methods are used to analyze the data such as Descriptive analysis, univariate, multivariate analysis, multiple regression analysis, linear regression analysis and structure equation model (SEM) which are as:

4.1 Descriptive Analysis

Using methods of summarizing and tabulating data that make their key properties more apparent, descriptive statistical analysis is used to explain the characteristics from empirical data.

4.2 Linear Regression Analysis

Regression analysis is a statistical method used to find the correlations between two or more variables that have cause-and-effect relationships and to utilize the correlation to create predictions about the subject. Exogenous and endogenous variables are the two types of variables employed in this analysis. The collected data were converted from the paper work to the computer (Microsoft Excel sheet) and represent it in a table format for further analysis. Due to the collection of comprehensive data set such as categorical and continuous, multivariate and linear regression analysis are used. The general equation for the linear regression is constructed below (refer Eq. 2).

$$Y = \alpha + \beta_1 X_1 + \beta_2 X_2 + \ldots + \beta_n X_n + \varepsilon t \tag{2}$$

whereas,

Y = dependent variable (Health).

α = Constant.

β = beta (Change/variation in independent variable).

X = Independent variable.

εt = Error term.

Multivariate analysis deals with more than one variable while, for prediction of identifying groups, cluster analysis will be used. The link between one dependent variable and one or more independent variables is explained by these regression estimations.

5 Analysis Results and Discussions

5.1 Time-Use and ATP

This research mainly focuses on the effect of ATP and built environment on physical and social health. Regarding time-use and travel participation, individual spend different amount of time while engage in a several activities such as out-of-home and in-home discretionary, mandatory, leisure, and maintenance activities by using diverse transport modes as shown in the Fig. 2. On weekdays and weekends, in-home obligatory was utilized more frequently by the person than out-of-home mandatory. However, individual spend more time on weekends in in-home mandatory as compared to weekdays which shows that most of the time on weekends individual spend it on their home and don't visiting to such a recreational places. In addition, individual perform much out-of-home activities on weekdays as compared to weekends which mean individuals spend their time on working or studies at weekdays and spend time at home on weekends.

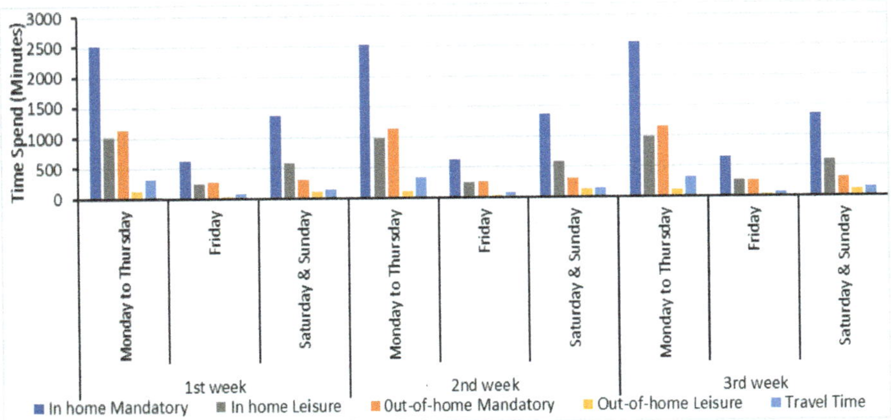

Fig. 2. Time spend for several activities in weeks

Furthermore, there is no such variation in out-of-home and in-home maintenance activities throughout the week while individual perform more in-home and out-of-home leisure activities on weekends as compared to weekdays. Regarding travel time, individual spend more travel time on weekdays by using different modes of transport which represents that individual perform mandatory activities on weekdays such as going to work, pick up/drop kids to school etc. The duration of travel over the week did not show any notable difference. Meanwhile, it was shown that responders often travel longest on weekdays. As seen in Fig. 3, Friday is the busiest day for both the total number of trips and the number of trip chains that begin and terminate at homes.

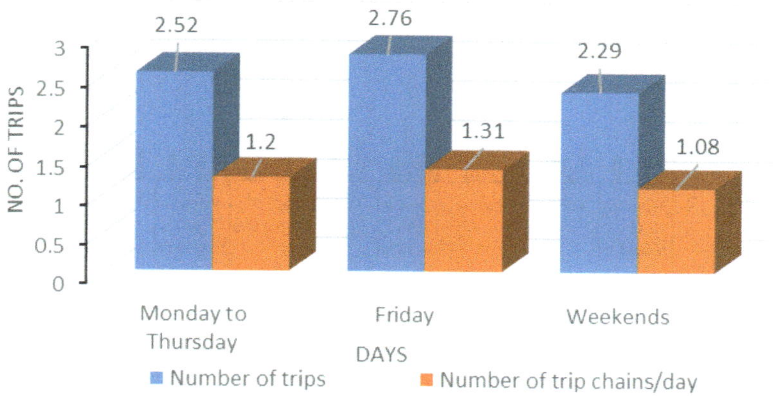

Fig. 3. Trip engagement and travel time spend

On weekends, there are less numbers of trip as well as trips chains which means that individuals engage in more in-home activities on weekends as compared to weekdays and Friday. However, at Friday individual perform maximum/highest number of trip and trip chain by visiting to Mosque or any other places and come back to the original destination (home) at the same day by using different modes of transport.

5.2 Travel Time and Transport Mode

In terms of travel time and transport modes, most of the time individual use motorized mode of transport for participating in different set of activities across the week as compared to other modes such as public transport and non-motorized mode of transport. There is a slight decrease in the public transport on weekends but have a slight increase in other modes (motorized and non-motorized). From the travel time it can be seen that as compared to weekends, individuals using more transport on weekdays across the week which represents that they are performing more out-of-home mandatory, discretionary, leisure as well as maintenance activities on weekdays and engage with more in-home activities on weekends as shown in Fig. 4 and Fig. 2 (time spend for several activities on a specified day across the week). Moreover, there is an abruptly increase in non-motorized transport mode on weekends and sharply decrease in other modes of transport as well in travel time.

In term of activity categories both the out-of-home and in-home activities are divided into leisure and maintenance activities time allocation. There is a pattern of activities throughout the week based on activity classification. Throughout the weekend, the allocation of in-home activities for all obligatory, leisure, and maintenance activities increased steadily in comparison to out-of-home required activities. In addition, there was a significant drop in out-of-home mandatory activities in time allocation however, there was a continuous increase in out-of-home maintenance and leisure activities as shown in table and Fig. 5. Moreover, there was a quiet little change in activity allocation across the weekdays.

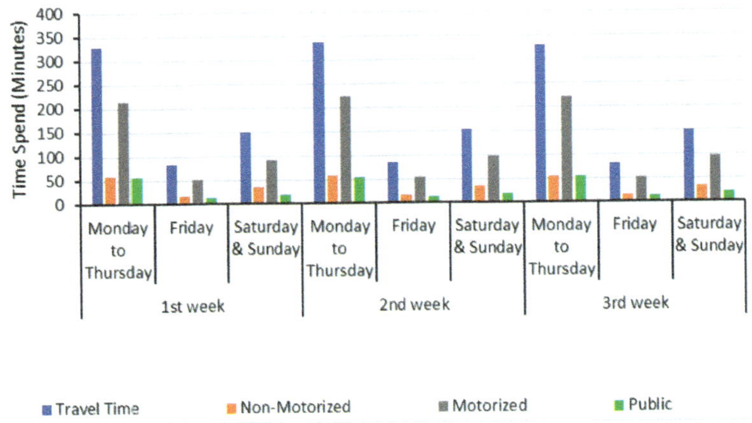

Fig. 4. Travel time and transport modes across the weeks

Table 2. Time spend for different activities

Characteristics	Days (Min)	Percentage/mean
In-home mandatory activities	Monday to Thursday	686.33 (46.66%)
	Friday	700.00 (48.61%)
	Weekends	738.81 (51.26%)
In-home leisure, maintenance activities	Monday to Thursday	302.65 (21.02%)
	Friday	313.81 (21.79%)
	Weekends	363.09 (25.21%)
Out-of-home mandatory activities	Monday to Thursday	319.73 (22.20%)
	Friday	277.96 (19.30%)
	Weekends	161.99 (11.25%)
Out-of-home leisure and maintenance	Monday to Thursday	57.23 (3.98%)
	Friday	72.46 (5.03%)
	Weekends	107.40 (7.46%)

Some on the activities are at a walkable distance whereas some of the activities are far from the individual home location. While participating in different set of out-of-home and in-home mandatory and discretionary activities, individual using several transport modes on different days throughout the week. Individual using more motorized mode of transport on weekdays as compared to Friday and weekends it's because of individual participating more out-of-home mandatory activities such as going to office, pick up/drop kids as compared to discretionary, leisure and maintenance activities. The participation of different set of activities can be seen in Fig. 5 whereas, travel time by different modes of transport can be seen in Fig. 6.

The percentage in the bracket displays the typical amount of time spent on travel and indoor and outdoor activities.

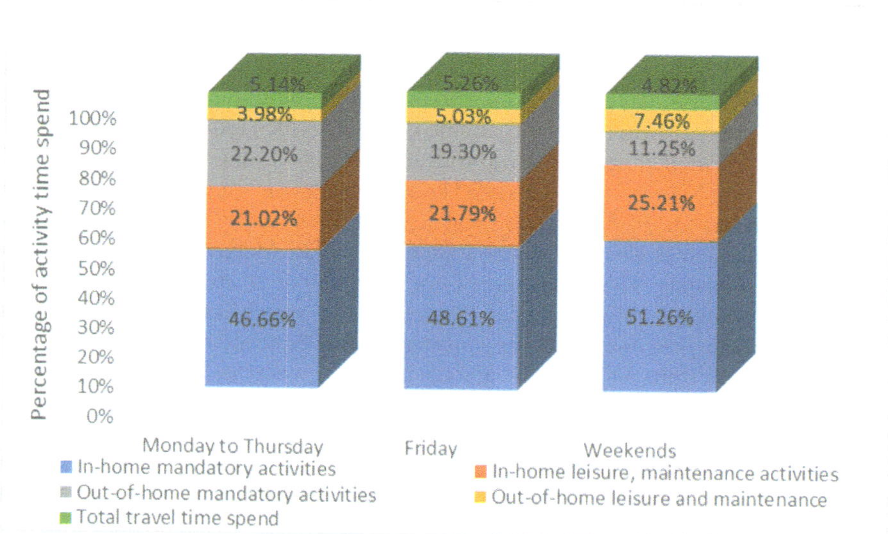

Fig. 5. Individual activity-travel time use

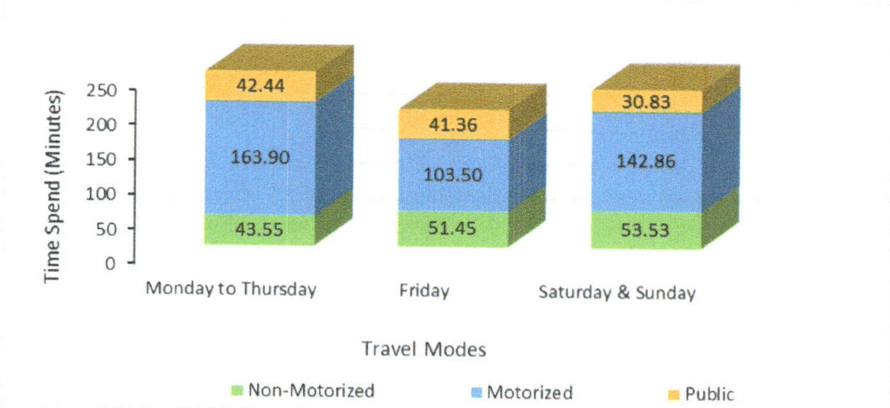

Fig. 6. Travel time by different travel modes

Overall it can be seen that, in-home mandatory, leisure and maintenance activities are significantly increasing while out-of-home mandatory activities are decreasing whereas the leisure and maintenance activities are increasing across the week. Most of the people using motorized mode of transport for their daily activity participation as compared to non-motorized and public transport. By using motorized mode of transport individual save their time but influencing physical activities which directly affect their physical health.

5.3 Frequency and Duration of Physical Activity

Physical activities is the movement of individual from one point to another or from place to place which requires the energy to expend. Intensity refer to the rate or magnitude of efforts required to perform an activity. Some of the activities required more energy to spend which are called as strenuous/vigorous intensity physical activity whereas, some of the activity require less energy to expend know as moderate intensity physical activity but overall, we can consider it as a physical activity. These activities are differentiating by various units such as we can measure it on minutes/day, Metabolic Equivalent Tasks (METs) and heartbeats. The intensities of our daily activities are based on the METs which is effective to our daily experience. Those activities which have intensity more than equal to 150 min/days or 3–6 METs are consider as a moderate intensity physical activity which require less efforts and burn less amount of energy whereas, those activities which have intensities more than 75 min/day or requires more energy to expand or greater than 6 METs are consider as strenuous/vigorous intensity physical activity.

The total amount of physical activity is a function of its intensity, frequency and duration. Frequency refer to how many days in a week individual perform an activity and can be measure in days/week whereas, duration raises to how long in a day individual perform an activity and are measured in minutes/day. Figure 8 shows the frequency of physical activity with the socio-demographic. As compared to female, male doing more both strenuous and moderate intensity physical activity whereas, in term of ages which are categorized as 15–22 years, 23–45 years, 46–55 years and above 55 years old. Those who are in the age of 23–45 years are performing high frequency (days/week) strenuous and moderate intensity physical activity compared to other ages. However, as compared to non-workers and students the workers are undertaking more vigorous and moderate physical activity as shown in the Fig. 7.

In addition the duration (minutes/day) of physical activity for male are more as compared to female whereas, individuals under the age of 23–45 years undertaking high duration of both types of physical activity with rest of the ages. Moreover, workers spend more time duration in day in different set of activities with high and moderate intensity than non-workers and students. However, those from high income undertaking more duration on strenuous physical activity whereas, individuals who are from low to middle income household performing more duration of moderate physical activity as compared to strenuous physical activity as show in the Fig. 8. According to vigorous intensity physical activity, individual participating in more moderate intensity physical activity in terms of days/week as well as minutes/day. Those who are from the age 15–22 years and low to middle income engage with more duration of moderate intensity physical activity as compared to vigorous physical activity.

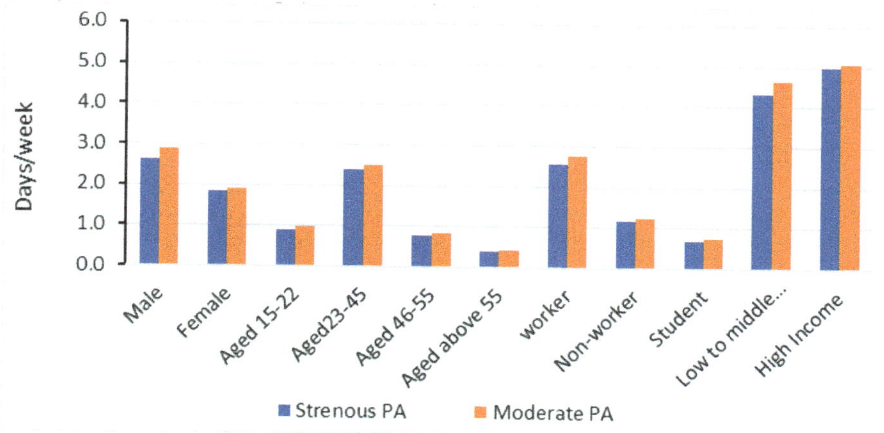

Fig. 7. Frequency of physical activity with socio-demographic

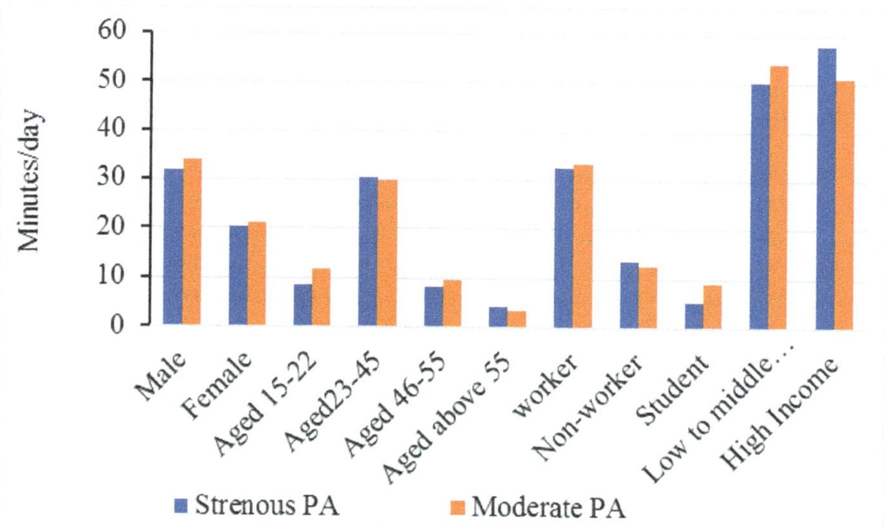

Fig. 8. Duration of physical activity with socio-demographic

5.4 The Proposed Model

The linear regression analysis will be performed on SPSS to find out the influence of time-use and activity travel participation and built environment on the physical and social health of travelers. As health and activity are inter related with each other sometimes activity influence heath whereas, health influence activity. As health directly influence activity whereas activity influence health through some intermediate variables. Different research used different intermediate variables to find out the influence of activity on health. In this research, different modes of transport and their physical intensity are used to find out the interaction between time use and ATP and built environment on the physical and social health as shown in Fig. 9.

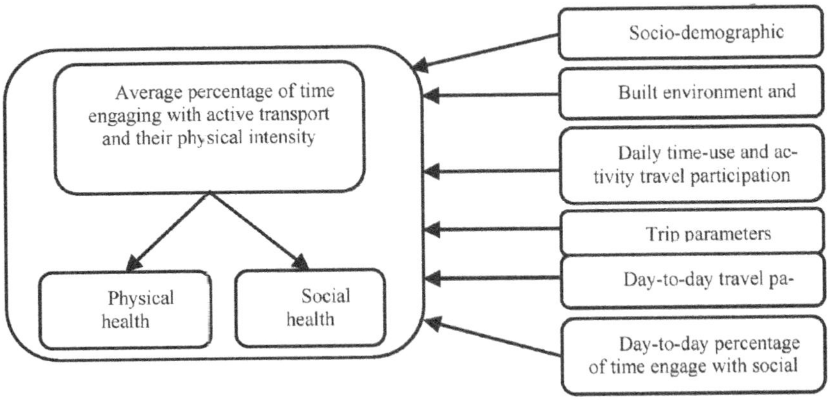

Fig. 9. The proposed theoretical mode

6 Conclusion

This study investigates the interaction of time-use and ATP, socio-demographic, trip engagement and travel-time spend for different set of activities at individual level. There is a significant increase in the in-home mandatory, maintenance and leisure activities throughout the week. Most the individuals are doing in-home mandatory on both weekdays and weekends which shows that individuals work for some days in a week/month. Individuals have high percentage of low household income, it's because of low education level (25.31%). The non-worker is in-between students and workers and the percentage of low income rate is highest (75.39%), whereas the percentage of middle and high income rate is very low because most of the individuals are workers and low education level. On the other hand individuals doing more out-of-home mandatory activities on weekdays as compared to Friday and weekends which represents individuals going to school or office on weekdays. Regarding trip participation, there was no discernible difference in the amount of time spent traveling each week. In the meanwhile, it turned out that responders often travel the most on Friday. In a similar vein, Friday sees the largest volume of both single journeys and chains of trips that begin and conclude at home. In addition, individuals spend more travel time on Friday and perform more trips and trips chains and less travel time on weekdays and weekends which tell us individuals going to masjid and perform more out-of-home maintenance and leisure activities time spend by travel.

References

1. Miller, H.J.: A measurement theory for time geography. Geogr. Anal. **37**(1), 17–45 (2005). https://doi.org/10.1111/j.1538-4632.2005.00575.x

2. Miller, H.J.: Modelling accessibility using space-time prism concepts within geographical information systems. Inter. J. Geographical Inform. Syst. **5**(3), 287–301 (1991). https://doi.org/10.1080/02693799108927856
3. Corbett, J.: Torsten Hägerstrand: time geography, vol. 12(7) (2005). doi: https://is.muni.cz/el/1431/jaro2006/Z0147/time_geography.pdf
4. Neutens, T., Schwanen, T., Witlox, F.: The prism of everyday life: towards a new research agenda for time geography. Transp. Rev. **31**(1), 25–47 (2011). https://doi.org/10.1080/01441647.2010.484153
5. Fauzi, N.F.M., Dharmowijoyo, D.B.: Activity-travel participation, multitasking in travel and daily well-being. In: MATEC Web of Conferences, vol. 270, p. 03014. EDP Sciences (2019). https://doi.org/10.1051/matecconf/201927003014
6. Macioszek, E., Sierpiński, G., Czapkowski, L.: Problems and issues with running the cycle traffic through the roundabouts. In: Transport Systems Telematics: 10th Conference, TST 2010, Katowice–Ustroń, Poland, 20–23 October 2010, pp. 107–114. Springer (2010). https://doi.org/10.1007/978-3-642-16472-9_11
7. Ali, M., et al.: The influence of COVID-19-induced daily activities on health parameters—a case study in Malaysia. Sustainability **13**, 7465 (2021). https://doi.org/10.3390/su13137465
8. Susilo, Y.O., Liu, C.: The influence of parents' travel patterns, perceptions and residential self-selectivity to their children travel mode shares. Transportation **43**(2), 357–378 (2016). https://doi.org/10.1007/s11116-015-9579-0
9. Pereira, A.: Live and let live: healthcare is a fundamental human right. Connecticut Public Interest Law J. **3**, 481 (2003)
10. Ali, M., Dharmowijoyo, D.B.E., de Azevedo, A.R.G., Fediuk, R., Ahmad, H., Salah, B.: Time-use and spatio-temporal variables influence on physical activity intensity, physical and social health of travelers. Sustainability **13**(21), 12226 (2021). https://doi.org/10.3390/su132112226
11. Macioszek, E.: Analysis of the effect of congestion in the Lanes at the inlet to the two-Lane roundabout on traffic capacity of the Inlet. In: Mikulski, J. (ed.) TST 2013. CCIS, vol. 395, pp. 97–104. Springer, Heidelberg (2013). https://doi.org/10.1007/978-3-642-41647-7_13
12. Axhausen, K.W.: Concepts of travel behaviour research. In: Threats from Car Traffic to the Quality of Urban Life: Problems, pp. 165–185. Emerald Group Publishing Limited, Causes and Solutions (2007)
13. Bhat, C.R., Koppelman, F.S.: Activity-based modeling of travel demand. In: Handbook of transportation Science, pp. 35–61. Springer, MA, Boston (1999)
14. Macioszek, E.: Geometrical determinants of car equivalents for heavy vehicles crossing circular intersections. In: Telematics in the Transport Environment: 12th International Conference on Transport Systems Telematics, TST 2012, Katowice-Ustroń, Poland, 10–13 October 2012. Selected Papers 12, pp. 221–228, Springer (2012). https://doi.org/10.1007/978-3-642-34050-5_25
15. Acker, V.V., Van Wee, B., Witlox, F.: When transport geography meets social psychology: toward a conceptual model of travel behaviour. Transp. Rev. **30**(2), 219–240 (2010). https://doi.org/10.1080/01441640902943453
16. Macioszek, E.: The application of HCM 2010 in the determination of capacity of traffic lanes at turbo roundabout entries. Transport Problems **11**(3), 77–89 (2016)
17. Wee, B.V., Ettema, D.: Travel behaviour and health: a conceptual model and research agenda. J. Trans. Health **3**(3), 240–248 (2016). https://doi.org/10.1016/j.jth.2016.07.003
18. Macioszek, E.: The comparison of models for follow-up headway at roundabouts. In: Recent Advances in Traffic Engineering for Transport Networks and Systems: 14th Scientific and Technical Conference" Transport Systems. Theory & Practice 2017" Selected Papers, pp. 16–26. Springer (2018). https://doi.org/10.1007/978-3-319-64084-6_2

19. Schwanen, T., Kwan, M.-P., Ren, F.: How fixed is fixed? Gendered rigidity of space–time constraints and geographies of everyday activities. Geoforum **39**(6), 2109–2121 (2008). https://doi.org/10.1016/j.geoforum.2008.09.002
20. Cullen, I., Godson, V.: Urban networks: the structure of activity patterns. Prog. Plan. **4**, 1–96 (1975). https://doi.org/10.1016/0305-9006(75)90006-9
21. Dharmowijoyo, D.B., Susilo, Y.O., Karlström, A., Adiredja, L.S.: Collecting a multi-dimensional three-weeks household time-use and activity diary in the Bandung Metropolitan Area, Indonesia. Trans. Res. Part A: Policy Pract. **80**, 231–246 (2013). https://doi.org/10.1016/j.tra.2015.08.001
22. Ali, M., Dharmowijoyo, D.B., Harahap, I.S., Puri, A., Tanjung, L.E.: Travel behaviour and health: interaction of activity-travel pattern, travel parameter and physical intensity. Solid State Technol. **63**(6), 4026–4039 (2020)
23. Macioszek, E., Iwanowicz, D.: A back-of-queue model of a signal-controlled intersection approach developed based on analysis of vehicle driver behavior. Energies **14**(4), 1204 (2021)
24. Petersen, J., Austin, D., Mattek, N., Kaye, J.: Time out-of-home and cognitive, physical, and emotional wellbeing of older adults: a longitudinal mixed effects model. PLoS One 10(10), Art no. e0139643 (2015). https://doi.org/10.1371/journal.pone.0139643
25. Ramli, M.I., Dharmowijoyo, D.E.: Activity-travel behaviour, and daily and global well-being. In: MATEC Web of Conferences, vol. 203 p. 05004. EDP Sciences (2018). https://doi.org/10.1051/matecconf/201820305004
26. Susilo, Y.O., Dijst, M.: How far is too far? travel time ratios for activity participation in the Netherlands. Trans. Res. Record: J. Trans. Res. Board **2134**(1), 89–98 (2009). https://doi.org/10.3141/2134-11
27. Macioszek, E., Sierpiński, G.: Charging stations for electric vehicles-current situation in Poland. In: Research and the Future of Telematics: 20th International Conference on Transport Systems Telematics, TST 2020, Kraków, Poland, 27–30 October 2020, Selected Papers 20, pp. 124–137. Springer (2020)
28. Farber, S., Páez, A.: My car, my friends, and me: a preliminary analysis of automobility and social activity participation. J. Transp. Geogr. **17**(3), 216–225 (2009). https://doi.org/10.1016/j.jtrangeo.2008.07.008
29. de Nazelle, A., et al.: Improving health through policies that promote active travel: A review of evidence to support integrated health impact assessment. Environment International **37**(4), 766–777 (2011). https://doi.org/10.1016/j.envint.2011.02.003
30. Dons, E., et al.: Transport mode choice and body mass index: Cross-sectional and longitudinal evidence from a European-wide study. Environ. Int. **119**, 109–116 (2018). https://doi.org/10.1016/j.envint.2018.06.023
31. Götschi, T., Hadden Loh, T.: Advancing project-scale health impact modeling for active transportation: s user survey and health impact calculation of 14 US trails. J. Trans. Health **4**, 334–347 (2017). https://doi.org/10.1016/j.jth.2017.01.005
32. Hodgson, S., Namdeo, A., Araujo-Soares, V., Pless-Mulloli, T.: Towards an interdisciplinary science of transport and health: a case study on school travel. J. Transp. Geogr. **21**, 70–79 (2012). https://doi.org/10.1016/j.jtrangeo.2012.01.011
33. Dharmowijoyo, D.B., Joewono, T.B.: Mobility and health: the interaction of activity-travel patterns, overall well-being, transport-related social exclusion on health parameters. In: Energy Efficiency in Mobility Systems, pp. 53–83. Springer (2020) . https://doi.org/10.1007/978-981-15-0102-9_4
34. Wee, B.V.: The traffic and transport system and effects on accessibility, the environment and safety: an introduction. In: The transport system and transport policy, Elgar, Cheltenham, pp. 4–18 (2013)

35. Zhang, J.: Urban forms and health promotion: An evaluation based on health-related QOL indicators. In: Proceeding of the 13th World Conference on Transportation Research, Rio de Janeiro, Brazil, 15–18 July, pp. 15–18 (2013)
36. Aghaabbasi, M., Ali, M., Jasiński, M., Leonowicz, Z., Novák, T.: On hyperparameter optimization of machine learning methods using a bayesian optimization algorithm to predict work travel mode choice. IEEE Access **11**, 19762–19774 (2023). https://doi.org/10.1109/ACCESS.2023.3247448
37. Chen, Y., et al.: Hybrid bayesian network models to investigate the impact of built environment experience before adulthood on students tolerable travel time to campus: towards sustainable commute behavior. Sustainability **14**(1) 325 (2022). https://www.mdpi.com/2071-1050/14/1/325
38. T. Ma et al.: Nonlinear relationships between vehicle ownership and household travel characteristics and built environment attributes in the us using the xgbt algorithm. Sustainability **14**(6), 3395 (2022). https://www.mdpi.com/2071-1050/14/6/3395
39. Mackett, R.L., Thoreau, R.: Transport, social exclusion and health. J. Transp. Health **2**(4), 610–617 (2015). https://doi.org/10.1016/j.jth.2015.07.006
40. Qian, Y., et al.: Classification of imbalanced travel mode choice to work data using adjustable SVM Model Appli. Sci. 11(24) 11916 (2021). https://www.mdpi.com/2076-3417/11/24/11916
41. Tang, P., Aghaabbasi, M., Ali, M., Jan, A., Mohamed, A.M., Mohamed, A.: How sustainable is people travel to reach public transit stations to go to work? a machine learning approach to reveal complex relationships. Sustainability **14**(7), 3989 (2022). https://www.mdpi.com/2071-1050/14/7/3989
42. Tao W., et al.: An advanced machine learning approach to predicting pedestrian fatality caused by road crashes: a step toward sustainable pedestrian safety. Sustainability 14(4), 2436 (2022). https://www.mdpi.com/2071-1050/14/4/2436
43. Xu, Z. Aghaabbasi, M., Ali, M., Macioszek, E.: Targeting sustainable transportation development: the support vector machine and the bayesian optimization algorithm for classifying household vehicle ownership. Sustainability **14**(17) 11094 (2022). https://www.mdpi.com/2071-1050/14/17/11094
44. Yang, L., et al.: Comparative analysis of the optimized KNN, SVM, and rnsemble dt models using bayesian optimization for predicting pedestrian fatalities: an advance towards realizing the sustainable safety of pedestrians. Sustainability **14**(17), 10467 (2022). https://www.mdpi.com/2071-1050/14/17/10467
45. Zhu, L., Ali, M., Macioszek, E., Aghaabbasi, M., Jan, A.: Approaching sustainable bike-sharing development: a systematic review of the influence of built environment features on bike-sharing ridership. Sustainability **14**(10) 5795 (2022). https://www.mdpi.com/2071-1050/14/10/5795
46. Raglin, J.S.: Exercise and mental health. Sports Med. **9**(6), 323–329 (1990). https://doi.org/10.2165/00007256-199009060-00001
47. Ainsworth, B.E., et al.: 2011 Compendium of physical activities: a second update of codes and MET values. Med. Sci. Sports Exerc. **43**(8), 1575–1581 (2011)
48. Dharmowijoyo, D.B., Susilo, Y.O., Syabri, I.: Time use and spatial influence on transport-related social exclusion, and mental and social health. Travel Behav. Soc. **21**, 24–36 (2020). https://doi.org/10.1016/j.tbs.2020.05.006

Levels of Evidence in Transport Research

Ireneusz Celiński(✉) and Grzegorz Sierpiński

Faculty of Transport and Aviation Engineering, Silesian University of Technology, Katowice, Poland

{ireneusz.celinski,grzegorz.sierpinski}@polsl.pl

Abstract. This article is an attempt to analyze the issue of the level of scientific evidence presented in broadly understood transport research. Various classifications of levels of evidence (LOE) have been formalized in the scientific literature and are reviewed in this article. They help assess the importance (rank) and quality of publicly presented scientific studies. On their basis, the recipient of scientific content forms an opinion on its usefulness and credibility. Not all LOEs are used in transportation studies. Those that are used are used with different frequencies. In this article, the authors analyze the problem of why more and less popular methods can be observed in scientific research in transport. The article tries to explain why transport research is currently being conducted in this way and, above all, whether it is possible to change the observed state of affairs in this matter.

Keywords: level of evidence · traffic research · GoogleScholar

1 Introduction

The classification of the level of scientific evidence refers to the research method used and described in a specific publication (taking into account publications from a specific discipline published in any publicly available sources). It is debatable whether expert opinions should be included in such a classification, because it is an opinion and not evidence or research [1–10]. If so, it includes: the opinion of the expert(s) (expert opinion), case report (case report/study), cross-sectional study (cross-sectional survey), case- control study study), cohort studies (cohort study), randomized controlled trial (randomized control trail), systematic review (systematic review), meta-analysis (meta- analysis) [11, 12]. The mentioned types of scientific research are often divided into subtypes, which are not the subject of interest in this article.

The credibility of the types of research presented above varies for their recipients [11, 12]. In general, it can be said that it increases from the level of expert opinion to systematic reviews and meta-analyses, along with the quantity and sometimes the quality of the research material collected. But is the more the better? Let us assume for now that this is in fact the case, assuming this LOE gradation. The increase in credibility by iterating through the types of methods mentioned above is by no means linear, much less obvious. This is due to the limitations of some of the methods.

G. Sierpiński et al. (Eds.): TSTP 2023, LNNS 907, pp. 82–101, 2024.
https://doi.org/10.1007/978-3-031-53181-1_6

A separate problem, somewhat beyond the above classification, is the reconstruction of historical research conducted under known assumptions in controlled conditions. Such verification of research conducted previously and usually by someone else allows us to determine to what extent this type of research (their results) can be repeated, and therefore to independently verify their reliability. An excellent negative example of this problem is an attempt to repeat cold fusion experiments [13].

An expert's opinion is one of the most popular forms of "proving", and at the same time it is burdened with the greatest errors, forecast errors [1–10]. Such an opinion can be formulated by any expert with any authority in the topic it concerns (professor, specialist engineer, psychologist, court expert). But even any person who is in any way involved in the problem, as long as he is considered an expert. Different categories of experts have been deliberately provided to indicate different ways of legitimizing opinions (in position, academic degree, function performed or assigned, etc.). Assuming substantive, domain-specific preparation on the part of the expert, the main disadvantage of this approach is the socio-economic basis of the opinion, which often affects the expert's perspective on the problem and thus the opinion issued [4, 7–9]. In today's turbulent socio-economic environment, it is often difficult to identify various dependencies affecting an expert. Each expert is part of one or more systems.

In turn, case studies/reports were one of the first types of applied research that gained great popularity in science and beyond [14, 15]. Hence, they currently constitute a large part of research, mainly in the humanities and, to a slightly lesser extent, in technical sciences. Case studies are particularly eagerly used in the social sciences, where the approach to the problem is oriented towards the practical use of knowledge (psychology, administration, sociology, transport, management, army, etc.).

Perhaps a more popular form of research than case studies is cross-sectional research. Very roughly, they can be reduced to survey research. They are a type of observational research. Respondents/participants in a cross-sectional study are selected on the basis of specific criteria (including or not including a given object from the population in the sample). What distinguishes these tests from others is that they are cheap and generally not time-consuming. Often, these studies are a preparation for more advanced forms of problem identification (e.g. cohort studies). In this type of research, it is difficult to determine causal relationships [16].

The case-control study consists of retrospection and observation. The aim of the study is to identify the factor inducing a given process/phenomenon. For this purpose, two groups are compared, one affected by the phenomenon/process and the other not. In research, data is collected primarily through interviews or analysis of reliable data (e.g. medical records). Interviews are unreliable due to the conditions of individual perception (respondent memory). The data collected from both groups (or, in fact, the differences) may indicate potential risk factors. Based on the research conducted, specific risk groups can also be identified. A case-control study, typical of medicine, examines the relationship between a phenomenon/process and a potential risk factor. The problem in these studies is to eliminate numerous errors: selection (time, identification), interview, history reconstruction (respondent memory) [16].

Cohort studies monitor numerous groups (cohorts) of subjects, usually over a longer period of time, even several decades. The time is counted from exposure/start of research until the effect occurs/end of research. The study is carried out in controlled conditions, often divided into groups, taking into account the specificity of the tested factor (distance covered on foot, number of cigarettes smoked, liters of alcohol consumed, etc.). According to many researchers, properly conducted cohort studies provide the best scientific evidence, but they require the involvement of significant resources. These studies make it possible to establish connections between exposure and effects [17]. Long research time can be an advantage but also a huge disadvantage when conditions change.

Clinical trials (typical of medicine), although they have a long history (over 2.5 thousand years), initiated a new era in medicine relatively recently. They are often called randomized control trials. These are active research related to interference in living organisms (in a biological sense) and carrying out targeted activities on groups of selected objects. The study includes large groups, consisting of groups or even thousands of objects. The group is divided randomly (hence randomization). One group is treated with experimental therapy (intentional interference in the body, e.g. with a new pharmacological agent) and the other is treated conventionally (typical/symptomatic treatment: for example aspirin) or is treated based on the placebo effect. After the therapy/intervention, both groups are compared. These studies must meet numerous conditions: the use of a large, statistically uniform sample, a double-blind study (neither of the parties carrying out the physical aspects of the study or undergoing therapy is informed about the type of therapy/intervention). The study should demonstrate differences between groups after the experiment/therapy is completed [18]. Due to the economic leverage they are associated with, these studies should be under special supervision of independent institutions.

Now it should be noted that tests typically associated with medical issues are rarely used in transport. Why? Costs are one side of the coin. The second is the appropriateness of using the equivalent of therapy in the transport network. In transport, such an implementation would be the equivalent of simultaneously conducted research and implementation. However, according to the authors, this problem requires detailed consideration in future research.

Systematic reviews are another type of research that uses repeatable analytical methods to collect secondary data and analyze it [19]. Using these methods, data is synthesized taking into account differences in the collected data (divergent research results on the same topic). Often, a systematic review does not include a meta-analysis [19]. The method is used to evaluate scientific research qualitatively or quantitatively. It can be said that these are summaries of the current state of knowledge. Evidence relevant to a given research question [19]. An important advantage here is the extensiveness of the accumulated knowledge base.

Meta-analysis is a quantitative approach that uses measures of the strength of the relationships of the variables included in the analysis. In a meta-analysis, the emphasis is on results from multiple studies as opposed to results from a single study [20].

Although many of the studies mentioned above involve verification of results (e.g. statistical control), external verification can also be distinguished. Verification consisting in repeating/reproducing historical research under the same, controlled conditions. Especially where analytical models and methods have been formalized in the domain knowledge base, this is a very interesting approach. This is especially the case in questionable cases, such as the previously mentioned cold fusion [13]. Research results positively verified by subsequent research teams are becoming more and more reliable, although Popper's principle of falsification should not be forgotten. Absolutely everything, but it is all true until it is falsified, apart from revealed truths.

1.1 Notes on the LOEs Used

An expert opinion can always be formulated if there is an expert willing to formulate it (especially if he receives remuneration for it). Assuming good substantive preparation of the expert, such opinions can be formulated by the expert in relation to repeatable phenomena known to him and verified by his professional experience. Each new instance of a case should aim at a different type of level of scientific evidence than an opinion or case study [21, 22]. Moreover, a significant obstacle to this type of arbitrary actions are the socio-economic conditions in which a given expert operates [23]. An interesting alternative is the use of the Delphi method, which can minimize these threats by eliminating outlier opinions. On the other hand, it is difficult not to see in this method some form of influencing and influencing opinions. The more crowded the market segment covered by the research, the more rigorously these opinions should be assessed.

A case study generally concerns one occurrence of some phenomenon/process. Although, sometimes several related events (similar in some sense) are analyzed. Therefore, the interest of this type of research is to describe, in essence, incidents that have occurred for the first time or have been described for the first time. This specificity of the case study means that in fact individual cases should not be compared. They are generally characterized by individual conditions. Therefore, in many scientific disciplines it is difficult to collect a set of statistically significant "similar" cases. Cases similar enough to each other. Only such a set of similar cases allows the formulation of scientific theories, in basic sciences. The more complex the organism, the less useful a case study appears to be. Transport networks are complex systems. This dependency is well demonstrated by unsuccessful implementations of ITS (Intelligent Transport Systems). In one case of the road network, the implementation will be successful, in another it may not be successful.

Another problem is the way of describing a given case. If the description is insufficient, it may lead to further comparisons of objects that are in fact significantly different from each other. Given the high complexity of the examined objects, this is very difficult (in road networks). Moreover, case studies are often used to covertly promote solutions or ideas.

Cross-sectional studies are carried out on a deliberately selected group, which is one of the main problems of these studies. A decline in interest in participating in various forms of surveys and interviews has been observed for years [24, 25]. The data obtained is as good as the questions were formulated correctly, including in an understandable way, and the answers were given on the topic and in accordance with the actual situation (an important problem, especially in medical interviews or other sensitive aspects of life). How good is a survey group at a time when undertaking free activities is not at all popular, and it is not volunteering after all. This type of research is becoming more and more popular, due to the development of social media, and is becoming more and more unreliable [26]. Many respondents either confuse simple facts, those that happened even relatively recently, or even confabulate. Finally, the barrier remains the price, in times when the basic criterion of almost every tender is the price, the lower it is, the worse representativeness we should expect from the survey (if such a thing exists at all). Many people are increasingly reluctant to engage in public research.

Case-control studies that compare two groups, one affected by some phenomenon/process and the other a control group, are carried out by analyzing the history and discovering differences in both groups. There is a problem of correlation and causality testing here, as purely statistical co-occurrence is often observed. Although this is a method that allows for isolating risk factors and focusing on the appropriate part of the population, it requires a lot of attention. The division into groups itself may be highly unreliable in certain cases of analysis. Further errors result from errors collected in historical documentation (e.g. the problem of digitization of medical and other records).

Cohort studies involve tracking a large number of deliberately selected subjects and monitoring them over a sufficiently long period. The study is conducted in controlled conditions, divided into target groups and allows for the observation of coincidences. The main problem here is obtaining a huge sample, available over a long period of time (decades), and maintaining appropriate diligence in preparing the results in such research conditions (Big Data issues). First of all, the research time is critical, as in many cases it does not allow revealing the nature of the studied phenomena and the dependencies occurring in the studied group. In transport, certain changes, e.g. the distribution of streams, not in terms of volume, but in terms of spatial distribution, may change for years. Communication behavior is similar.

Randomized control trials where groups are divided randomly (randomization) may cause errors already at this stage. This is due to poor understanding of stochastic processes even among researchers (sampling procedure, use of specialized randomness tests e.g. DieHard stat test). Therefore, further research with incorrect division may be questioned. Therefore, maintaining further conditions, such as sample size, statistical homogeneity, and double-blind testing, does not guarantee success. The phrase "the study must show differences between the groups after the end of the experiment" is often found in the literature, which is interesting in itself and indicates that the study is focused in terms of the result. The test usually has to do something if someone paid for it. For this reason, this type of research is also falsified. This forgery also generally requires payment.

A systematic review is a physical compilation of results previously published in a certain subject and in a specific field of literature. It allows us to reveal differences between different studies on the same problem. But the problem here is that unpublished research or research published in limited publications (in specific language) is not included in the synthesis. What about publications in less popular languages or dialects? It is often said that combining research in this way creates a new (in the sense of larger) data set. In such large samples, statistical anomalies are smoothed out. However, with data incompatibilities, this is difficult. Such aggregation of research also allows for the elimination of the risk of unusual coincidences. It also allows you to take into account less intuitive statistical effects. The purpose of such a review should not be forgotten, as it may influence the method of aggregation and selection of data. At the end, there is always a risk that "something" has been omitted. If any important data is omitted, it calls into question the entire review. And what about significantly outlier cases? Throw them away or save them?

Verification of research previously conducted by someone, although appropriate, is generally carried out from a different reference level, at a different time, in different conditions and may serve purposes other than purely scientific ones. Such a goal, in the conditions of strong competition among research centers, may be the desire to undermine someone's achievements. Therefore, it can often be implemented with other assumptions, including "non-scientific ones". Apart from basic research, where methods and models are transparent, in humanities and sometimes technical fields, this can often lead to proving only the right truth. As an aside, some studies are sometimes not worth looking at all.

It is always worth considering creating initiatives involving sharing basic data in the cloud, open to everyone. We leave the conclusions for consideration and the data for use by others.

1.2 Possibilities and Limitations of Using LOE in Transport

The following describes how the different levels of scientific evidence apply in practice and how they can be applied to transport. Some of them are commonly used in transport, some are not, especially those directly identified with, for example, medical science. Moreover, as it will be shown later in the text of the article, this is completely unnecessary due to the fact that transport is often compared to a living organism, and the analogy is by no means unfounded. It is commonly known that transport is the lifeblood of the economy. The transport network is born like a person and like a person grows old. He also has similar problems in childhood, adolescence and adulthood. The analogy is all too clear.

Expert opinion, also broadly understood expert methods, are widely used in transport. In fact, in every area of transport, such opinions are expressed [27–30]. What is worth emphasizing in this context is that the area of transport is becoming more and more specialized, which means that expert specializations are also emerging: electromobility, sustainable development, traffic control, traffic organization, traffic safety, autonomous vehicles, etc. Some of them overlap in many aspects, such as electromobility and sustainable development, some not, such as traffic control, which is a relatively narrow specialization (similarly as autonomous vehicles). Therefore, you should pay attention to the competences and experience of the people who form the opinions. Moreover, in transport, contradictory strategic tendencies are increasingly emerging and may influence expert opinions, for example: paid vs. free public transport, combustion vehicles vs. electric vehicles, full traffic control vs. free traffic, etc. To sum up, everyone has an opinion, especially in transport. Which everyone knows, especially from the perspective of the user of the transport network, which is extremely flat (perspective). It is worse if the expert's perspective is also individual.

The case study is extremely popular as a research method in the field of transport. Virtually every element of the transport network, point and linear, as well as their aggregates, every geographical area, every specific transport industry, can be the subject of a case study [31–33]. There are even dedicated journals for the purpose of publishing case studies [33]. Unfortunately, there is one fundamental problem. The larger the fragment of the transport network described in such a study, the lower its reference value for the remaining parts of the network. Specialists in the field of road traffic engineering know how large the dynamics of changes can be in one, even in a short section of the road network. In road traffic, only certain units can be compared. This type of approach may be helpful in terms of an example of the application of some technical or organizational solution in transport, but never as a basis for generalization. Case studies are often created in the context of large linear and point projects implemented in the transport network. On the other hand, transport systems operating on Big Data sets, which will be able to collect very detailed information, may contribute to increasing the role of case studies in transport research by increasing the comparability of various transport systems, by very precise parameterization in time and space.

Cross-sectional research is the bread and butter of transportation systems, historically and today. Even more so now, when ITS and passenger information systems have become widespread. These systems help collect Big Data in the area of transport, in near real time, without the need to conduct time-consuming and expensive face-to-face interviews [34–39]. These tests become cheaper every year, and their processing becomes faster and easier. The only threat in this area seems to be two phenomena. The first is the anonymization of social life by protecting basic data (which complicates the analysis of communication behavior), and the second is the threat of reinterpretation of the obtained results in the interest of their owner or the institution financing transport (e.g. data on the filling of PTZ funds).

A case-control study searched for transport references in WWW sources no longer brings as many hits (links) as previous types of studies [40–43]. The vast majority of views on scientific literature concern medical cases. And those that include the term: case-control study (the results of the WWW searches presented in this article were carried out in English) concern primarily issues related to traffic safety, where, as in medical research, there is a human factor, and the condition of individuals in the observed sample is often analyzed from a specific transport point of view (some form of interdisciplinary research, generally assigned to the field of medicine). Therefore, it is still a medical problem but studied in the context of transport. Although this is a typically medical test, nothing prevents it from being used in transport research. The large sample size is problematic and requires the involvement of a large number of resources and time. However, with the current sensor networks present in the road system, this is not a problem. C-ITS (cooperative ITS) systems will be introduced soon! The fundamental issue is the division into research groups, which requires extensive experience in the organization and operation of road networks.

In terms of cohort studies, the situation is much better in terms of the number of scientific articles relating to transport issues [44–50]. But even in this case, most of the mentioned studies, even though they are conducted in the field of transport, still concern, in one way or another, health aspects and/or health protection.

In principle, this type of research should be obligatory in transport due to the long period of infrastructure operation and well-defined factors disturbing its efficient functioning. Specific elements of point and rope infrastructure are exposed to specific hazardous factors. For example, a large number of accidents involving frontal collisions occur at intersections (as opposed to roundabouts).

Similar to case-control studies, the same applies to randomized control studies [51–54]. Here, too, the number of "transport" items is negligible, and they concern very different issues, often directly related to medical issues, and only the location of the facilities is typically transport-related.

As you can see, research associated primarily with medicine (clinical, control, randomized) is rarely used in transport in the *perse sense*. Is this justified? It seems not, because each of the procedures of randomized trials, starting from the randomization of groups to the use of therapy (including changes in organization, traffic control at selected facilities) can be used in relation to the transport network. Double blinding may be problematic. They can be implemented, for example, by transferring competences in the field of monitoring and installation/modernization of infrastructure and equipment outside the local authority. All this means that this type of research in transport may generate additional, very high costs, but it is still possible to implement it. However, in this way we will exclude local impacts in the transport network.

Systematic reviews are very popular in transport issues [55–63]. The issues covered cover almost all modes of transport and many different practical aspects of transport. The reservations that should be made in this matter are similar to those in the case study, the issue concerns the method and validity of data aggregation, which are treated as "the same" or "similar". Any set of data not included in the synthesis may become the basis for the falsification of theories built through the synthesis of research materials from various centers. In the last 10 years, 2022–2012, in one of the leading transport journals, Transportation Research Part B: Methodological, the number of publications has doubled (data from Scupus.com, respectively: 619, 359 articles).

Another problem is that the reviews take into account accepted research directions in the so-called mainstream. Research that sheds some new light on mainstream research is often overlooked. In traffic engineering, this is the concept of traffic capacity, which has become the holy grail in the study of transportation systems. Meanwhile, research by consulting companies shows that in Polish cities over 30% of intersections operate beyond the capacity threshold, often even quite efficiently (e.g. Szczecin). Were the thresholds or methodological bases defined incorrectly?

Meta-analysis in transport issues is as popular as other "non-typically medical" records of scientific evidence [64–70]. The issues discussed here also concern a wide range of transport issues. Are meta-analyses valid in transport? It is safe to say that this will depend on the area of research. Certain general or more precise conclusions drawn on the basis of research on the "entirety" of publications in a given area may be problematic, e.g. for road traffic engineering due to the variety of infrastructure solutions used, means of transport, traffic rules, etc. For railway traffic engineering, such activities may be much more useful due to the conditions of this mode of transport. Similarly, meta-analyses will be useful in the field of migration research, air traffic, maritime transport, and passenger transport. However, they may also be unreliable in relation to: tariff systems, organization of mass events, traffic modelers, etc., to name just some of the problematic areas.

1.3 Lack of Research

An interesting issue in the context of this article is the question: "What if there is no research? What research? Because they are different as shown above. Their availability varies. What research to start with? This is where systematic reviews and meta-analyses can help. For this purpose, one should look for similar problems described using these levels of command by analogy. To do this, you should follow the procedure described later in this article and verify which methods are most often effective in a given issue. What is important here is the context of the research, the so-called was the study verified in the context of a systematic review - positive or negative? Can we count it as representative in a given issue?

1.4 What Dose of Therapy?

A very difficult problem in the context of transport and randomized trials is choosing something that corresponds to the amount of therapy in these trials. In medicine, there are specific doses of preparations both in classic treatment and for new therapeutic procedures (apart from variants of the dose itself).

In randomized transport studies, the problem is how to select infrastructural, organizational and traffic control changes so that comparison in such a study makes sense and involves some rational costs. In the context of systematic reviews and meta-analyses, the question is how to compare studies where different therapies were used?

In transport, there are various research sites where unusual activities were carried out in some parts of the network to test some new forms of organization and control. The problematic thing is that it generally took place on single objects, in limited road routes, etc. There are Dutch attempts to check how the lack of any regulations in the field of traffic organization affects road traffic safety. There are known Swedish attempts at solutions aimed at improving the safety of vulnerable road users, etc. In fact, it has always been a problem of research costs. But these limitations are eliminated by collaborative design within the European Union. This type of financing enables control activities and proposed transport therapy to be carried out in various study groups. In this context, another question arises: how many objects are subjected to a specific "transport" therapy? The same applies to the problem of selecting the number of objects to be examined and dividing them into groups. How many objects should be analyzed? For example, in two neighboring voivodeships in Poland we have ca. 57,000 PTZ stops. How to select them for this type of controlled and randomized studies? What should be treated as therapies in the context of this number and in relation to the type and purpose of the facility? However, the number of objects in transport systems, which often reaches thousands and hundreds of thousands, certainly favors the implementation of this type of research. So why are they not used (see data in tables at the end of the article)?

1.5 How to Compare Objects?

As mentioned above, when discussing systematic reviews of "intersections are not equal to intersections", the same applies to linear sections of road infrastructure. Each of such objects has hundreds of attributes at a high level of detail. This is compounded by the dynamics of the traffic stream or streams. In road traffic, this is an unregulated, highly random process. The parameters of the traffic stream change in different cross-sections, sometimes every dozen or so meters (e.g. traffic density).

Therefore, the basic problem is the selection of reasonable aggregates for comparing various objects in the road network. But in medicine, different cases are compared, even though each organism is very different. Therefore, it is worth introducing similar taxonomies in transport as in medicine.

2 Statistical Research on LEO in Transport

2.1 Assumptions

In the analysis of the frequency of occurrence of various types of research, it was decided to use popular websites related to scientific bibliometrics and typical and recently popular terms related to transport issues.

The following websites were used in the statistical review: Elsevier, Google Scholar, Springer and Taylor and Francis [71–74]. In addition, the following general and recently popular terms were used, such as: *transport, transport network, traffic flow, traffic control, ITS, autonomous veh., traffic safety, electromobility, smart cities (free choice of article authors).* The conducted study omitted expert opinions. The most popular medical terms were also excluded from the study due to the fact that in the case of controlled and randomized studies they would result in an overrepresentation of exclusively medical articles due to the frequent use of transport terms in medical articles, e.g. transport, transport system, movement, etc. In the Fig. 1 presents an example query in the Google Scholar search engine.

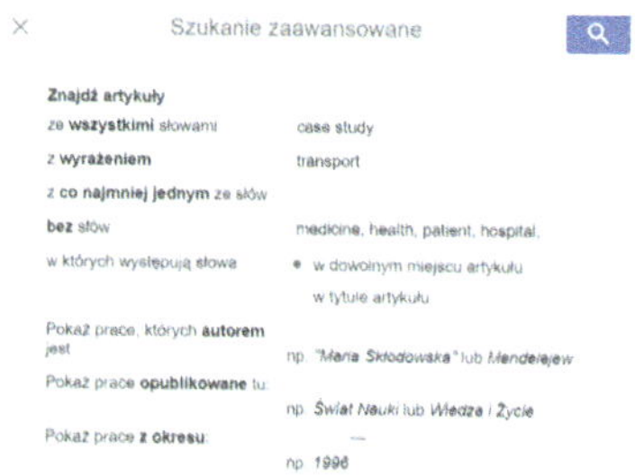

Fig. 1. Example search query in Google Scholar

Tables 1, 2, 3, 4, 5, and 6 present the results of searching databases of popular websites for thematic articles related to terms occurring in the field of transport, especially those recently popular such as electromobility or smart city. The choice of terms is solely the author's.

2.2 Query Results

The tables present the results of queries carried out in various search engines on scientific and similar websites: Springer, Taylor-Francis, Google Scholar. Data obtained from Google Scholar as the website most open to public access are discussed in detail.

Table 1. Review by transport term in Elsevier Pub.

LOE/term	CSR	CSS	CCS	CS	RCT	SR	MA
Transport	25,275	16,993	22,146	22,354	17,056	20,932	18,633
Transport Network	25,216	12,686	21,725	24,469	19,204	23,504	20,654
Traffic flow	25,514	13,165	22,129	22,714	17,043	21,262	18,321
Traffic Control	29,080	19,073	19,998	26,691	14,391	26,082	23,331
ITS	36,654	30,446	29,857	30,729	26,398	20,127	27,865
Autonomous veh	23,513	9,054	20,206	20,420	15,147	29,833	16,890
Traffic safety	25,216	12,421	21,698	22,563	16,515	20,574	18,446
Electromobility	23,109	9,518	19,835	19,936	14,168	17,696	15,755
Smart cities	23,908	10,902	20,551	20,857	15,302	18,890	16,844

CSR (case report/study), CSS (cross-sectional survey), CCS (case-control study), CS (cohort study), RCT (randomized control trail), SR (systematic review), MA (meta- analysis)

Table 2 shows the results for the same queries obtained in the Google Scholar search engine. It is the only scientific data archiving system with such open availability. Over uses one of the best search engines for web resources. This data, unlike the search engines of journals such as Springer or Elsevier, has a global reach, hence the greater number of hits for a specific query presented in Table 2. Therefore, only this list was analyzed, the rest were compiled only for statistical purposes.

Table 2. Review by transport term in Google Scholar [–].

LOE/term	CSR	CSS	CCS	CS	RCT	SR	MA
Transport	3,140,000	214,000	3,580,000	52,300	50,300	1,320,000	1,460,000
Transport Network	67,000	2,670	52,700	602	15,600	13,000	10,300
Traffic flow	42	4,870	238000	968	19,500	26,200	17,800
Traffic Control	253,000	3,730	281,000	1,040	17,500	24,700	19,400
ITS	31,900	453	25,400	172	11,600	14,900	3,840
Autonomous veh	57,300	998	58,000	343	16,400	9,440	7,690
Traffic safety	76,600	3,450	62,600	842	15,800	13,400	9,760
Electromobility	7,700	140	5,950	135	303	1,920	2,750
Smart cities	99,100	1,730	71,300	653	16,300	20,700	15,400

Table 2 shows the number of articles found for particular terms in the context of particular types of scientific evidence records. At first glance, there is a large diversity of results obtained, both for individual terms and for subsequent methods used in research. In this context, we can talk about popular research methods (used often) and niche research methods (used rarely). The problem considered in this article is whether or not the popularity of a given method is substantively justified, which will be described later in the article.

Table 3. Aggregate review by transport term in Google Scholar [%].

LOE	CSR	CSS	CCS	CS	RCT	SR	MA	Catfish
%	32.31	2.01	37.87	0.49	1.41	12.50	13.39	100

Table 3 presents the aggregates from Table 2 in the form of the percentage of publications for which a specific type of study was used. Therefore, based on the analysis of the data in Table 2, it can be concluded that the most popular types of research are case studies, control studies (>30%) and, at a similar level, meta-analyses and systematic reviews (approx. 13%). Cohort, randomized and cross-sectional studies are the least popular (<2%).

Is this distribution of the data presented in Table 2 justified? According to the authors, not with the current development of technology in transport systems. With the use of ITS, ATCS (advanced traffic control systems) and SIP (passenger information systems) systems, the least popular cohort, randomized and cross-sectional studies should be used much more often. Perhaps this is due to the fact that researchers approach these methods as typically medical procedures.

Table 4. Review by transport term in Google Scholar [%].

LOE	CSR	CSS	CCS	CS	RCT	SR	MA	Sum
Transport	31.99	2.18	36.47	0.53	0.51	13.45	14.87	100
Transport Network	41.39	1.65	32.56	0.37	9.64	8.03	6.36	100
Traffic flow	0.01	1.58	77.43	0.31	6.34	8.52	5.79	100
Traffic Control	42.14	0.62	46.80	0.17	2.91	4.11	3.23	100
ITS	36.14	0.51	28.78	0.19	13,14	16.88	4.35	100
Autonomous veh	38.16	0.66	38.62	0.23	10.92	6.29	5.12	100
Traffic safety	41.98	1.89	34.31	0.46	8.66	7.34	5.35	100
Electromobility	40.75	0.74	31.48	0.71	1.60	10.16	14.55	100
Smart cities	44.01	0.77	31.66	0.29	7.24	9.19	6.84	100

Table 4 shows the percentage distribution of data on individual test terms and methods, based on Table 2. This table highlights even greater disproportions in the applicability of individual methods in transport. They are particularly visible in the area of issues related to smart cities, road traffic safety and autonomous vehicles. The "niche nature" of cohort and cross-sectional studies is particularly visible in this review, and in the latter case it is extremely puzzling. What is noteworthy is the very small number of case studies in the area of traffic flow analysis compared to other methods. With regard to systematic reviews and meta-analyses, there is a growing interest in electromobility, autonomous vehicles and smart cities.

Therefore, based on the data analysis, it can be concluded that the most popular types of research for the term transport are case studies, control studies and, at a similar level, meta-analyses and reviews. The least popular are cohort, randomized and cross-sectional studies. That is, exactly as for all transport terms taken together.

Therefore, based on the analysis of the data in table…, it can be concluded that the most popular type of research for the term transport network are definitely case studies and control studies. Randomized studies come next, followed by meta-analyses and syntheses. Cohort and cross-sectional studies are the least popular.

Therefore, based on the analysis of the data in the Tables 2 and 3. It can be concluded that the most popular type of research for the term traffic flow is definitely control research. Case study research is the least popular.

Therefore, based on the data analysis, it can be concluded that the most popular type of research for the term transport control are definitely case studies and control studies. Cohort studies are the least popular. Similarly, it can be said that the most popular type of research for the term ITS is definitely case studies. Cohort studies are the least popular. The most popular type of research for the term autonomus vehicle are definitely case studies, and follow-up studies. Cohort studies are the least popular. The most popular types of research for the term traffic safety are definitely case studies and control studies. Cohort studies are the least popular. The most popular type of research for the term electromobility are definitely case studies and control studies. The least popular are cross-sectional and cohort studies. The most popular type of research for the term smart cities are definitely case studies and control studies. Cohort and cross-sectional studies are the least popular.

Table 5. Review by transport term in Springer Pub.

LOE/term	CSR	CSS	CCS	CS	RCT	SR	MA
Transport	179,947	114,221	20,683	25,851	27,189	37,109	54,570
Transport Network	107,164	41,845	8,307	10,653	11,088	21,221	25,015
Traffic flow	56,840	16,042	4,406	5,927	6,301	13,026	12,890
Traffic Control	80,398	25,140	8,333	10,192	10,459	18,219	21,486
ITS	29,040	8,014	1,481	2,052	2,396	8,057	6,284
Autonomous veh	30,675	7,513	1,728	2,267	2,728	8,228	6,678
Traffic safety	45,895	15,134	4,264	5,989	7,053	13,645	12,930
Electromobility	608	272	76	71	68	222	196
Smart cities	46,933	11,072	2,497	3,652	4,718	12,276	10,322

CSR (case report/study), CSS (cross-sectional survey), CCS (case-control study), CS (cohort study), RCT (randomized control trail), SR (systematic reviev), MA (meta-analysis)

Below are the answers to questions directed to the Springer publishing house's search engine. In this case, we did not limit ourselves to fields and disciplines by searching the entire database of articles. An extended (advanced) search engine was used, Fig. 2.

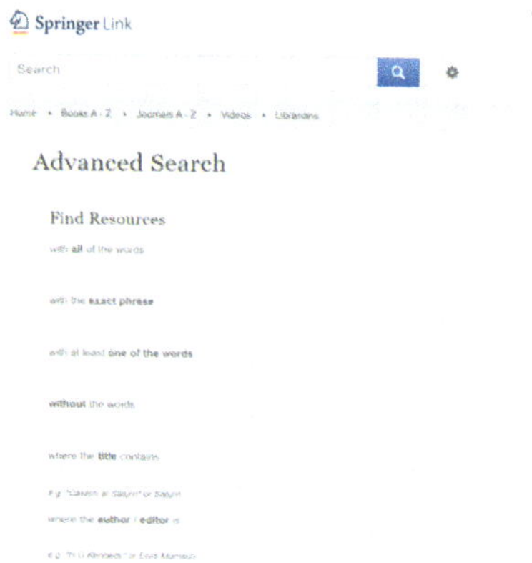

Fig. 2. Example search engine in Springer

The obtained results are presented in Table 5. When searching, be careful to leave spaces between the various search options.

Below are the answers to questions directed to the search engine of the Taylor&Francis publishing house (Table 6).

Table 6. Review by transport term in Taylor&Francis.

LOE/term	CSR	CSS	CCS	CS	RCT	SR	MA
Transport	562,174	44,111	427,641	38,502	68,766	342,464	32,288
Transport Network	198,397	13,604	159,675	14,229	25,985	139,051	13,361
Traffic flow	38,208	3,471	55,206	2,230	9,661	43,178	4,717
Traffic Control	65,056	8,016	136,118	9,645	23,346	87,667	8,955
ITS	36,052	3,030	57,637	3,430	10,193	48,272	4,377
Autonomous veh	30,727	2,531	49,985	4,376	7,220	39,788	4,452
Traffic safety	56,045	4,140	48,040	4,768	10,035	40,055	4,117
Electromobility	355	145	324	189	149	269	151
Smart cities	39,758	2,180	30,516	3,177	5,413	27,978	3,030

CSR (case report/study), CSS (cross-sectional survey), CCS (case-control study), CS (cohort study), RCT (randomized control trail), SR (systematic review), MA (meta-analysis)

3 Directions of Research in Transport

First of all, the structure of transport research broken down into individual specializations (traffic engineering, logistics, transport systems, life cycle analysis, reliability, etc.) results from the substantive preparation of research and scientific-teaching staff. Most scientists publish in their field, some of them publish articles in interdisciplinary teams. In Poland transport is in scientific discipline of: civil engineering, geodesy and transport. We can distinguish specialists focused on narrow topics and scientists who publish more broadly. This division is somewhat secondary to the university's offer addressed to students and doctoral students. Before the introduction of doctoral schools, the student-master system was still in force. Theoretically, this offer should be correlated with the market needs for specialists in specific fields. However, this is often not the case due to choices made in second and third level studies.

Another division of the focus of publicly published research results from what is happening on the market and the currently implemented design topics. Often, research not related to current market trends or financed research topics of projects becomes niche. Project activities in particular require open publication of research topics.

Such research as presented in this article also requires an analysis of the dynamics of phenomena. This will make the data more realistic in the field of currently niche areas of scientific activity in transport in the coming years. The distribution of published texts will change in the coming years, placing emphasis on contemporary research: electromobility, smart cities, C-ITS, artificial intelligence in transport, machine learning, etc. It is also possible to increase the representativeness of this type of publications by limiting the analysis time, which includes the article was not implemented due to the difficulty of indicating the exact time frame for issues emerging in the field of transport. Certain issues are developed over a longer period of time, and only the emergence of appropriate technologies and new developments allow them to be developed on a full scale. This is the case, for example, with machine learning or neural networks, which have been known since the last century and only after 2000 have their development accelerated rapidly. The research presented by the authors is as of today and does not take into account the dynamics of changes in publications.

4 Conclusions

The results presented in the tables above for individual publishing houses and organizations indicate that these entities use search engines and search mechanisms with different functionalities. The authors did not recognize the mechanisms in the publications and what criteria they use when searching for articles on a given topic. The authors' goal was to show the distribution of statistical weight describing individual publications in various groups of scientific research (LOE) and for various keywords. Selection of keywords: transport, transport network, traffic flow, traffic control, ITS, autonomous vehicle, traffic safety, electromobility, smart cities, is dictated by the subjective views of the authors of the article on the problem in question. Nevertheless, both general transport concepts and new issues analyzed within the field, such as electromobility, autonomous vehicles and smart cities, were taken into account.

An overview of the type presented in this article has its strengths and weaknesses. The presentation of results strongly depends on the formulated criteria for searching databases of scientific publications. There were no complicated inclusions or restrictions in the queries.

The results presented in this article clearly show that the interest of the authors of publications in particular fields varies. The low interest in issues such as electromobility, smart cities and autonomous vehicles may result from the fact that these are relatively new issues (young fields of research), so the number of publications in knowledge bases is small, but will systematically increase.

The presented data clearly show that research traditionally assigned to the field of medical sciences CCS (case-control study), CS (cohort study), RCT (randomized control trail) enjoy relatively little interest among transport researchers. The analysis of the titles of individual studies (e.g. Elsevier's search engine shows articles with covers and a short description of the content not presented in this article, a random review of authors) shows that these are texts, mainly in the field of medicine or on its borders (you can treat them as interdisciplinary) with other disciplines, in this case transport.

Such a study should be repeated in a few years to study the dynamics of the processes:

- development of new issues in the area of transport such as electromobility, smart cities and autonomous vehicles. Intuition indicates that unless there are drastic changes in transport conditions, these issues should be described more and more often and be more strongly represented both in publications and at industry conferences,
- investigate the use of typically medical research methods in transport in the following years, although the intuition cannot be maintained here as in the previous point. Such changes require a different look at the issue of transport, a different organization of research, in large research teams while maintaining high funding for this research, as is the case in the Horizon programs (EU programme),
- it would be reasonable to perform this type of review for isolated issues, then these distributions can be analyzed at a different level of detail.
- the authors are planning another article in which the area of randomized control trials will be subjected to such a detailed study, with a detailed review of each published article,
- the authors also plan to supplement this text based on a survey conducted among the Polish transport research community, which will include research into the reasons for using these and not other research methods.

The authors have made every effort to formulate queries in appropriate search engines, please remember that any typo results in incorrect results, which are not always so obvious. The entire review was conducted in English, which may significantly underestimate the presented results in some areas (in Poland there is a large group of transport magazines published only in the native language, although this number is systematically decreasing).

The overview presented in this article should be considered as a rough guide. Showing the distribution of the weight of articles between individual fields. To narrow it down too much would miss many articles. This often results in the data showing articles from subject areas other than transport, mainly medicine.

Table 7. Review by transport term in Google (in millions, rounded down)

LOE/term	CSR	CSS	CCS	CS	RCT	SR	MA
Transport	699	239	1150	120	116	174	161
Transport Network	619	508	818	80	234	243	346
Traffic flow	127	102	129	12	69	58	24
Traffic Control	387	370	862	36	177	107	81
ITS	239	1470	4320	10	1,120	199	43
Autonomous veh	55	247	333	2	76	32	9
Traffic safety	548	385	892	152	186	200	112
Electromobility	0.4	0.08	0.2	27	0.03	0.1	58
Smart cities	313	495	516	14	343	61	44

CSR (case report/study), CSS (cross-sectional survey), CCS (case-control study), CS (cohort study), RCT (randomized control trail), SR (systematic review), MA (meta-analysis)

The Table 7 presenting a query carried out in the Google search engine shows how transport terms are located in the general knowledge database. This is intended solely to demonstrate the general distribution of these term combinations. Here, the low results of recently top daily transport issues, such as autonomous vehicles or electromobility, are also noteworthy. However, this is probably still a relatively narrow research area.

This type of review may indicate currently fashionable research directions and studies that are beneficial from the point of view of bibliometric analytics.

References

1. Chande, P.K., Tokekar, S.V.: Expert-based maintenance: a study of its effectiveness. IEEE Trans. Reliab. **46**(1), 53–58 (1998)
2. Clemen, R.T., Winkler, R.L.: Combining probability distributions from experts in risk analysis. Risk Anal. **19**(2), 187–203 (1999)
3. Cooke, R.M.: Excalibur – Windows Version of Excalibur: Software for Performance Based Combination of Expert Judgments. Department of Mathematics, Delft Uni (2001)
4. Goossens, L.H.J., Harper, F.T., Kraan, B.C.P., Metivier, H.: Expert judgment for a probabilistic accident consequence uncertainty analysis. Radiat. Protect. Dosimet. **90**(3), 295–303 (2000)
5. Hatfield, A.J., Hipel, K.W.: Risk and systems theory. Risk Anal. **22**(6), 1043–1057 (2002)
6. Kline, A., Renn, O.: A new approach to risk evaluation and management: risk based, precaution-based, and discourse-based strategies. Risk Anal. **22**(6), 1071–1094 (2002)
7. Kraan, B.C.P., Cooke, R.M.: Processing expert judgment in accident consequence modeling. Appear. Radiat. Protect. Dosimet. (Exp. Judgm. Accident Conseq. Uncertain. Anal. Spec. Issu.) **90**(3), 311–315 (2000)
8. Myung, I.J., Ramamoorti, S., Bailey, A.D., Jr.: Maximum entropy aggregation of expert predictions. Manage. Sci. **42**(10), 1420–1436 (1996)
9. Skjong, R., Eknes, M.L.: Societal risk and societal benefits. Risk Decis. Policy **7**, 57–67 (2002)
10. West, M.:, Modeling Expert Opinion. Bayesian Statistics 3, pp. 493–508. Oxford University Press (1988)

11. Greenhalgh, T.: How to Read a Paper: The Basics of Evidence-Based Medicine, 5th edn. Wiley (2014)
12. Łamża, Ł.: Śwaty Równoległe, Czarna Owca. Wołowiec (2020)
13. Fleischmann, M., Pons, S.: Electrochemically Induced Nuclear Fusion of Deuterium. J. Electroanal. Chem. **261**, 301 (1989)
14. Rebolj, A.B.: The case study as a type of qualitative research. J. Contemp. Educ. Stud. 28–43 (2013)
15. Mills, A.J., Durepos, G., Wiebe, E. (eds.): Encyclopedia of Case Study Research. SAGE Publications, Inc. (2010). https://doi.org/10.4135/9781412957397
16. Setia, M.: Methodology series mod. 3: Crosssectional studies. Indian J. Dermatol. **61**, 261–264 (2016). https://doi.org/10.4103/0019-5154.182410. http://www.med.mcgill.ca/epidemiology/hanley/c609/material/CaseControlStudiesBreslow2007.pdf
17. Morrow, B.: An overview of cohort study designs and their advantages and disadvantages. Int. J. Therapy Rehabili. **17**, 458–463 (2010). https://doi.org/10.12968/ijtr.2010.17.10.78810
18. Stolberg, H., Norman, G., Trop, I.: Randomized controlled trials. AJR Am. J. Roentgenol. **183**, 1539–1544 (2005). https://doi.org/10.2214/ajr.183.6.01831539
19. Nunn, J., Chang, S.: What are systematic reviews?. Wiki J. Med. **7**, 5 (2020). https://doi.org/10.15347/WJM/2020.005
20. Shelby, L., Vaske, J.: Understanding meta-analysis: a review of the methodological literature. Leisure Sci. **30**, 96–110 (2008). https://doi.org/10.1080/01490400701881366
21. Kroger, J.:. The Applicability and Limitations of Expert System Shells **38** (1989)
22. https://www.ilearnlot.com/expert-system-advantages-disadvantages/34332/
23. https://www.oecd.org/corruption/anti-corruptionethicscompliancehandbook.pdf
24. Macik, R.: Ankiety internetowe w percepcji osób korzystających i niekorzystających z nich w pracy zawodowej [Web Surveys (Cawi) Perception by Marketing Professionals Classified as Cawi Users and Non-Users]. Studia Ekonomiczne/Uniwersytet Ekonomiczny w Katowicach. **195**, 125–139 (2014)
25. https://spis.gov.pl/
26. Biemer, P., Groves, R., Lyberg, L., Mathiowetz, N., Sudman, S.: Measurement errors in surveys. Technometrics **35** (1991). https://doi.org/10.2307/1269316
27. Urbański, M., Sudyka, J., Grondys, K.: Expert evaluation of road infrastructure management. **20**, 465–492 (2021). https://doi.org/10.7409/rabdim.021.028
28. Policy Options to Decarbonise Urban Passenger Transport Results of Expert Opinion Survey Guineng Chen, ITF/OECD 19 April 2018. https://www.itf-oecd.org/sites/default/files/docs/expert-opinion-survey-decarbonising-urban-passenger-transport.pdf
29. Chekmarevaa, G., Khamburovaa, D., Matyashchukb, S., Fillimonovac, E., Mamaev, E.: XIII International Conference on Transport Infrastructure: Territory Development and Sustainability Expert Methods for Assessing the Quality of Transportation Services.
30. Norhisham, S., Abu Bakar, M.F., Mohd Azlan, N.N.I., Apandi, S.N.D.I.Y., Khalid, N.H.N.: Overview of methodology for developing expert system based on expert view for quality of urban bus services in Peninsular Malaysia (2021): https://doi.org/10.6007/IJARBSS/v11-i12/11829
31. Ogryzek, M., Adamska-Kmieć, D., Klimach, A.: Sustainable transport: an efficient transportation network—case study. Sustainability **12**, 8274 (2020). https://doi.org/10.3390/su12198274
32. Singh, A., Gurtu, A., Singh, R.K.: Selection of sustainable transport system: a case study. Manag. Environ. Qual. **32**(1), 100–113 (2021). https://doi.org/10.1108/MEQ-03-2020-0059
33. Case Studies on Transport Policy a Journal of the World Conference on Transport Research Society. https://www.elsevier.com/journals/case-studies-on-transport-policy/2213-624X?generatepdf=true

34. Bonnel, P., Munizaga, M.: Transport survey methods - in the era of big data facing new and old challenges. Transp. Res. Procedia **32**, 1–15 (2018). https://doi.org/10.1016/j.trpro.2018.10.001
35. Eisenmann, C., Kuhnimhof, T.: Some pay much but many don't: vehicle TCO imputation in travel surveys. Transp. Res. Procedia **32**, 421–435 (2018). https://doi.org/10.1016/j.trpro.2018.10.056
36. Wielinski, G., Trépanier, M., Morency, C., Habib, K.N.: Comparing multiple data streams to assess free-floating carsharing use. Transp. Res. Procedia (2018)
37. Wittwe, R.R., Hubrich, S., Wittig, S., Gerike, R.: Development of a new method for household travel survey data harmonisation. Transp. Res. Procedia (2018)
38. Zijlstra, T., Wijgergangs, K., Hoogendoorn-Lanser, S.: Traditional and mobile devices in computer assisted web-interviews. Transp. Res. Procedia (2018)
39. Zmud, J., Lee-Gosselin, M., Munizaga, M., Carrasco, J.-A. (eds.): Transport Survey Methods: Best Practice for Decision Making, Emerald, UK (2013)
40. Shatu, F.M., Kamruzzaman, M.D.: Investigating the link between transit oriented development and sustainable travel behavior in Brisbane: a case-control study. J. Sustain. Develop. **7**(4), 61–70 (2014)
41. Connor, J., et al.: Driver sleepiness and risk of serious injury to car occupants: population based case control study. BMJ **324**(7346), 1125 (2002)
42. Allerston, J., Justham, D.: A case-control study of the transit times through an accident and emergency department of ankle injured patients assessed using the Ottawa Ankle Rules. Accid. Emerg. Nurs. **8**(3), 148–154 (2000)
43. Galéra, C., Orriols, L., M'Bailara, K., Laborey, M., Contrand, B., Ribéreau-Gayon, R.: Mind wandering and driving: responsibility case-control study. BMJ **345**, e8105 (2012)
44. Pyko, A., et al.: Long-term exposure to transportation noise in relation to development of obesity—a cohort study. Environ. Health Perspect. **125**(11), 117005 (2017)
45. Cantuaria, M.L., et al.: Residential exposure to transportation noise in Denmark and incidence of dementia: national cohort study. BMJ **374**, n1954 (2021)
46. Perchoux, C., Kestens, Y., Thomas, F., Van Hulst, A., Thierry, B., Chaix, B.: Assessing patterns of spatial behavior in health studies: their socio-demographic determinants and associations with transportation modes (the RECORD Cohort Study). Soc. Sci. Med. **119**, 64–73 (2014)
47. Saunders, L.E., Green, J.M., Petticrew, M.P., Steinbach, R., Roberts, H.: What are the health benefits of active travel? A systematic review of trials and cohort studies. PLoS ONE **8**(8), e69912 (2013)
48. Lee, W., Kang, M.Y., Kim, J., Lim, S.S., Yoon, J.H.: Cancer risk in road transportation workers: a national representative cohort study with 600,000 person-years of follow-up. Sci. Rep. **10**(1), 11331 (2020)
49. Brown, J.B., et al.: Geographic variation in outcome benefits of helicopter transport for trauma in the United States: a retrospective cohort study. Ann. Surg. **263**(2), 406 (2016)
50. Poulos, R.G., et al.: Near miss experiences of transport and recreational cyclists in New South Wales, Australia. Findings from a prospective cohort study. Accid. Anal. Prev. **101**, 143–153 (2017)
51. Goetz, H.M., Creutzinger, K.C., Kelton, D.F., Costa, J.H.C., Winder, C.B., Renaud, D.L.: A randomized controlled trial investigating the effect of transport duration and age at transport on surplus dairy calves: Part I. Impact on health and growth. J. Dairy Sci. **106**(4), 2784–2799 (2023)
52. Awad, L.N., Reisman, D.S., Pohlig, R.T., Binder-Macleod, S.A.: Reducing the cost of transport and increasing walking distance after stroke: a randomized controlled trial on fast locomotor training combined with functional electrical stimulation. Neurorehabil. Neural Repair **30**(7), 661–670 (2016)

53. Rosenfield, A., Attanucci, J.P., Zhao, J.: A randomized controlled trial in travel demand management. Transportation **47**(4), 1907–1932 (2020)
54. Brough, R., Freedman, M., Ho, D.E., Phillips, D.C.: Can transportation subsidies reduce failures to appear in criminal court? Evidence from a pilot randomized controlled trial. Econ. Lett. **216**, 110540 (2022)
55. Spasojevic, B., Lohmann, G., Scott, N.: Air transport and tourism–a systematic literature review (2000–2014). Curr. Issue Tour. **21**(9), 975–997 (2018)
56. Pournader, M., Shi, Y., Seuring, S., Koh, S.L.: Blockchain applications in supply chains, transport and logistics: a systematic review of the literature. Int. J. Prod. Res. **58**(7), 2063–2081 (2020)
57. Bellizzi, M.G., Eboli, L., Mazzulla, G.: Air transport service quality factors: a systematic literature review. Transp. Res. Procedia **45**, 218–225 (2020)
58. Fraser, S.D., Lock, K.: Cycling for transport and public health: a systematic review of the effect of the environment on cycling. Eur. J. Public Health **21**(6), 738–743 (2011)
59. Wanner, M., Götschi, T., Martin-Diener, E., Kahlmeier, S., Martin, B.W.: Active transport, physical activity, and body weight in adults: a systematic review. Am. J. Prev. Med. **42**(5), 493–502 (2012)
60. Aranda-Balboa, M.J., Huertas-Delgado, F.J., Herrador-Colmenero, M., Cardon, G., Chillón, P.: Parental barriers to active transport to school: a systematic review. Int. J. Public Health **65**, 87–98 (2020)
61. Larouche, R., Mammen, G., Rowe, D.A., Faulkner, G.: Effectiveness of active school transport interventions: a systematic review and update. BMC Public Health **18**, 1–18 (2018)
62. Villa-González, E., Barranco-Ruiz, Y., Evenson, K.R., Chillón, P.: Systematic review of interventions for promoting active school transport. Prev. Med. **111**, 115–134 (2018)
63. Heath, G.W., Brownson, R.C., Kruger, J., Miles, R., Powell, K.E., Ramsey, L.T.: The effectiveness of urban design and land use and transport policies and practices to increase physical activity: a systematic review. J. Phys. Act. Health **3**(s1), S55–S76 (2006)
64. Holmgren, J.: Meta-analysis of public transport demand. Transp. Res. Part A: Policy Pract. **41**(10), 1021–1035 (2007)
65. Zamparini, L., Reggiani, A.: Freight transport and the value of travel time savings: a meta-analysis of empirical studies. Transp. Rev. **27**(5), 621–636 (2007)
66. Melo, P.C., Graham, D.J., Brage-Ardao, R.: The productivity of transport infrastructure investment: a meta-analysis of empirical evidence. Reg. Sci. Urban Econ. **43**(5), 695–706 (2013)
67. Dimitropoulos, A., Oueslati, W., Sintek, C.: The rebound effect in road transport: a meta-analysis of empirical studies. Energy Economics **75**, 163–179 (2018)
68. Zamparini, L., Reggiani, A.: Meta-analysis and the value of travel time savings: a transatlantic perspective in passenger transport. Netw. Spat. Econ. **7**, 377–396 (2007)
69. Nijkamp, P., Pepping, G.: Meta-analysis for explaining the variance in public transport demand elasticities in Europe. J. of Transp. and Stat. **1**(1), 1–14 (1998)
70. Bastiaanssen, J., Johnson, D., Lucas, K.: Does transport help people to gain employment? A systematic review and meta-analysis of the empirical evidence. Transp. Rev. **40**(5), 607–628 (2020)
71. Elsevier Search. https://www.elsevier.com/search-results
72. Google Scholar. https://scholar.google.pl/
73. Springer. https://link.springer.com/
74. Taylor&Francis. https://taylorandfrancis.com/

Modern Transport Infrastructure as Part of Critical Infrastructures

Edyta Boratyńska-Karpiej(✉) and Paweł Engel

Department of Transport Analysis, Centre for EU Transport Projects, Warsaw, Poland
{eboratynska-karpiej,pengel}@cupt.gov.pl

Abstract. Transport infrastructure (roads, railway lines, navigable canals, seaports and airports) form the country's transport network and, due to their importance, are classified as critical infrastructure. As a result of various crisis events, this infrastructure may be destroyed, damaged, which will negatively affect the operation of the state. Therefore, it is necessary to constantly develop and expand the infrastructure, especially in the context of its protection. They serve, among others, projects financed from EU funds.

In the article, we will consider what critical infrastructure is and why transport-related infrastructure belongs to it. We will also diagnose potential threats that may negatively affect the functioning of the state. We will also provide examples of attacks on critical transport infrastructure, which have recently intensified. The reason for this is, among others, the geopolitical situation we find ourselves in. Finally, we will also provide opportunities to obtain funding to improve the operation of critical transport infrastructure. The source of funds may be EU funds.

Keywords: transport · critical infrastructure · transport system · infrastructure projects in the transport sector · financing · military mobility projects · EU funds

1 Introduction

At the beginning of our considerations, let us first try to explain what critical infrastructure is. The term is not very old and is certainly not universally recognised. It first appeared in the United States and Canada in the 1990s; it defined "systems and installations indispensable to operation of modern society and administration" [1].

It should be noted that in European Union (EU) Council Directive 2008/114/EC of 8 December 2008. (Article 2) [5]: "critical infrastructure means an asset, system or part of infrastructure located in the territory of the Member States which is essential for maintenance of vital societal functions, health, safety, security, material or social well-being of the population disruption or destruction of which would have a significant impact in a Member State as a result of loss of those functions". This is a slightly different approach to the subject then American ones. Taking this into account, in Poland, the concept was introduced by the **Crisis Management Act** [2], and defined as **"systems and their constituent functionally related objects, including buildings, equipment, installations, services that are key to the security of the state and its citizens and**

G. Sierpiński et al. (Eds.): TSTP 2023, LNNS 907, pp. 102–113, 2024.
https://doi.org/10.1007/978-3-031-53181-1_7

that serve to ensure the efficient functioning of public administration, as well as institutions and entrepreneurs".

The basic criterion for qualifying specific facilities as critical infrastructure is their impact on the security of the state and citizens. Detailed criteria are prepared by the Government Center for Security (GCS) in cooperation with the relevant ministers and central authorities responsible for that part of the infrastructure that can potentially be classified as critical. The entry into force of the Crisis Management Act created basic organized mechanisms for the protection of critical infrastructure and defined the systems included in it.

Critical infrastructure includes the following systems:

- energy, energy commodity and fuel supply,
- communications,
- ICT networks,
- financial,
- food supply,
- water supply,
- health protection,
- transport,
- rescue,
- systems that ensure continuity of public administration,
- production, storage, warehousing and use of chemical and radioactive substances, including pipelines for hazardous substances.

To present the components of the critical infrastructure system more vividly, we will use examples. The critical infrastructure system includes in particular:

- Transport infrastructure facilities: stations, airports, railway networks, river and sea ports, highways, bridges, tunnels and passenger service points (MOP);
- Power plants, transformer stations, traction substations, transmission networks and other energy infrastructure facilities;
- Water management companies dealing with the protection and use of available surface and groundwater resources, as well as pumping stations, water intakes, sewage treatment plants and other water supply and sewage infrastructure facilities;
- Telecommunications networks and masts, data storage and processing centers;
- Military bases and barracks, food and weapons warehouses, border guard posts and commands, plants producing equipment for the army;
- Hospitals, clinics, laboratories and emergency stations;
- Food producing plants and enterprises responsible for food distribution;
- Police stations and headquarters, prosecutor's offices, courts, prisons, detention centers;
- Drilling platforms, refineries, mines, gas plants, gas stations, fuel bases, gas ports, oil ports, cement plants, grain silos, wood processing plants, chemical and petrochemical plants and hazardous waste landfills;
- Financial facilities and institutions: banks, treasuries, mints, insurance companies;
- Government administration facilities, including offices and other facilities.

Many of the objects listed above are functionally interconnected, i.e. disruption of one of them affects the others. Therefore, solutions and systems aimed at ensuring the safety and continuity of operation of such facilities and installations constitute the basic element of their equipment.

Transport infrastructure (roads, railways, navigable canals, seaports and airports, petrol stations, etc.) form the country's transport network and, due to their importance to the national economy, they are classified as critical infrastructure. In addition, public transport facilities such as stations, bus stops, waiting rooms should also be indicated as critical infrastructure.

The essence of the activities related to critical infrastructures is not only to ensure their protection against hazards, but also to ensure that any damage or disruption to their functioning is as short-term as possible, easily remedied and does not cause additional losses to citizens and the economy.

As you can see, critical infrastructure is an extremely complex concept due to the diversity of facilities and services that constitute it and their constant evolution. Along with the development of infrastructure included in critical infrastructure, threats that may cause disruptions in the operation of this infrastructure change and develop. The aim of the article is to show how the concept of critical infrastructure is defined in applicable legal acts - both Polish and European - and what is the importance of critical infrastructure (including transport) for the proper functioning of the state. Due to the geopolitical situation we are in (especially after Russia's invasion of Ukraine), the article will present the basic threats causing disruptions in the operation of critical infrastructure and present examples of attacks on transport infrastructure. However, due to the continuous development of technology and human capabilities, it is impossible to clearly define all threats to this infrastructure. Therefore, at the end, the article indicates the need to constantly modernize and expand critical infrastructure, which investments can be financed from EU funds.

2 Hazards

Hazards to transport critical infrastructure can be divided like in Fig. 1.

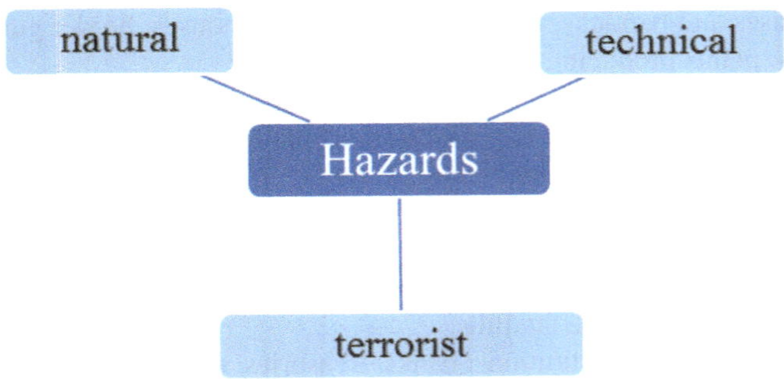

Fig. 1. Diagram - Hazards to transport critical infrastructure - own study

Other division of hazards to critical infrastructure, including transport, are [3]:

a) Hazards in the environmental area (natural disasters, unreliability of energy systems and back-up power supply, sabotage and terrorist threats to physical installations and facilities, intrusion resistance and integrity of access control systems).
b) Hazards to the technological area (absence of alternative transmission paths, occurrence of manufacturing or design faults, poor quality and security of software in use, resistance of transmission technology to interception or electromagnetic interference and environmental conditions, equipment lifetime to first failure and resistance to operating conditions).
c) Hazards to the data and network area (through transmission methods and data transfer protocols, design and scaling, standards for implementing equipment from different suppliers, lack of synchronisation of the network with the elements of the transmission path, degree of complexity).
d) Hazards from the humans (theft, sabotage, terrorism, revenge, disgruntled citizens or employees, unconscious malicious action due to user ignorance, lack of security policy and unrealistic regulations or loopholes, lack of qualified managers and lack of training and awareness campaign).

Natural hazards include all environmental hazards causing natural disasters, e.g. floods, heavy snowfall, icing, heavy winds, etc.

Hazards caused by natural forces impede operations of all transport systems and therefore affect functioning of critical infrastructure. They should therefore be taken into account as one of the most important factors affecting its efficient functioning. Floods in Poland (including the most widespread one in 1997) are an example of the negative impact of nature. Consequences for critical infrastructure include those related to:

- energy, raw material and fuel supply system;
- food supply system - destruction of crops;
- transport infrastructure - damaged roads, railways and bridges; damage to telecommunications infrastructure;
- production facilities - stoppage of operations;
- hydrotechnical structures - disruption or damage to dikes.

Technical failures affecting critical infrastructure in transport also occur frequently - it is the damage or destruction of a structure or a system of technical equipment that causes an interruption or destruction of its use, which has occurred in a sudden and unforeseen manner. This happens, for example, on the railways.

Terrorism is one of the most important hazards of the contemporary geopolitical situation. Any of the systems (including transport) classified as critical infrastructure can be the target of a terrorist attack. Critical infrastructure related to transport and communication, on the other hand, is one of the main targets of terrorist attacks. Each mode of transport and communication is vulnerable: air, rail, road and maritime. Very vulnerable facilities are airports, aeroplanes, seaports, railway stations and trains, including subways, where terrorists can use bomb charges and chemical weapons.

With the current development of technology, critical infrastructure is also very vulnerable to cyber-attacks. There are a growing number of examples of this in Poland, for example at the beginning of this year, passengers on urban transport in the Upper Silesian and Zagłębie Metropolitan Area lost access to an application that allows them to pay for travel on trams, buses and trolleybuses belonging to the Metropolitan Transport Authority. The card readers of the Silesian Public Services Card stopped working in the vehicles and the information boards at the stops. This was a ransomware attack, involving the encryption of server disks.

Non-functioning traffic control systems, non-functioning ticket machines and limited access to the passenger information system. These, however, are just some of the effects of the hacking attack that took place on the night of 24–25 June 2023 on the Olsztyn transport system. More than two weeks have passed since these events and the system was still not functioning, moreover, paper tickets have also gone missing. This shows how large and long-lasting the consequences of such attacks can be.

Rail

Despite changes in the so-called modal split, rail is still an important tool for moving both people and goods. At the same time, despite the passage of time, it is still vulnerable to attacks. The robbers with guns in their hands are replaced by hackers whose weapon is a mouse. The power of the threat of cybercriminal attacks was realized by, among others, Danes. Hackers hit all trains of DSB, a public operator in Denmark, stopping them for several hours. The attack was aimed at the supplier of the software used in DSB. An application used by railway workers to control speed and events on the tracks was infected with ransomware. Supeo immediately suspended all programs. However, these are the basic working tools of railway workers. As a result, no trains ran on the route, and those running were stopped.

Poland also recently had a series of "strange" railway events. On line 273, the Daleszewo - Szczecin Główny section, and on line 351, the Choszczno - Szczecin Główny section, there was a hack on the frequency of the Polish railways and an unauthorized transmission of a radio stop signal, which causes an emergency stop of the trains. The effect of this event was difficulties in train running - on one day alone, as a result of an unauthorized radio stop signal, over 20 trains were delayed. The situation repeated itself on other lines in the following days.

Neither Denmark nor Poland are exceptions. In 2022, attacks were also directed towards German or Italian railways. At the end of March, there was a cyber-attack on the Italian state railway operator - Ferrovie dello Stato Italiane (FS). An intrusion into the IT systems of two FS daughter companies was detected - Trenitalia (responsible for train operation) and Rete Ferroviaria Italiana (manager of Italian railways). In order to thoroughly check the systems and eliminate faults, FS preventively suspended ticket sales at its ticket offices and from ticket machines at railway stations, while online sales continued normally. There are many indications that it was a ransomware attack, but there is no information about the possible consequences for the company or whether any data was stolen. FS stated that train traffic itself was not interrupted, many transport companies complained about long delays. The Czech railway operator Metrans that all its trains had been stopped since the afternoon of March 23 due to an attack on the IT systems of the Italian railway infrastructure. Several other carriers also reported train stoppages and delays.

Russia's allies cannot feel safe either. The Belarusian Railways have already been attacked several times by hackers (also called cyber guerrillas). The aim of the attack was to disrupt the deployment of troops and fuel supplies to Russian and Belarusian military units. The railway had to switch to manual control, which will significantly hamper logistics and transport in Belarus. However, what is important, it will not cause emergency situations, so as not to lead to a possible disaster in rail traffic.

This shows that no one can feel safe. A cyber threat can be as catastrophic as a physical attack and can be directed against any country.

It appears that protective measures are being taken due to the seriousness of the situation. Allied Market Research in the report "Cybersecurity Market Global Opportunity Analysis and Industry Forecast, 2021–2031" forecasts that EU regulations, technological development, as well as the war in Ukraine and the related increase in the threat of cyberattacks will trigger an investment impulse in the implementation of transport security technologies rail. The cybersecurity market in the railway industry, currently worth USD 9.77 billion, will almost double by 2029, reaching USD 16.77 billion. Expenditures will increase on the protection of passenger data, ensuring the security of systems monitoring the traction condition and managing sidings, as well as on the modernization of the security of passenger and freight train traffic control systems.

3 Critical Infrastructure Protection

Critical infrastructure protection should be understood as all activities aiming at ensuring functionality, continuity and integrity of critical infrastructures in order to deter, mitigate and neutralise threats, risks or vulnerabilities, and to recover rapidly in case of failure, attacks or other events which compromise their proper operations. At the same time, it should be borne in mind that the protection of critical infrastructure should be comprehensive, which means that the following areas should be taken into account in the organisation of protection: physical, technical, personal, ICT and legal protection, as well as the assistance of the government party during its reconstruction.

3.1 European Union

European critical actors are increasingly interconnected and interdependent, making them stronger and more effective, but also more vulnerable in the event of an incident. Russia's war against Ukraine has brought with it new threats - both physical and in the form of cyberattacks - often occurring together in a hybrid form. The sabotage of the Nord Stream gas pipelines and other recent incidents have clearly demonstrated that the resilience of EU critical infrastructure is at risk. There is an urgent need to strengthen the EU's capacity to protect against attacks on critical infrastructure, both within its own territory and in its immediate vicinity.

Already in 2022, in response to the geopolitical situation, the European Commission presented a proposal to increase the resilience of EU critical infrastructure. The proposal for a Council Recommendation was based on a five-point plan for the resilience of critical infrastructure. The recommendation aims to intensify and accelerate work to protect critical infrastructure in three priority areas: preparedness, response and international cooperation. This will be achieved through stronger support and coordination from the Commission to improve preparedness and response to current threats. The Recommendation also envisages deepening cooperation between Member States and with neighboring third countries. Priority should be given to the key sectors of energy, digital infrastructure, transport and space.The EU has a specific role to play when it comes to infrastructure that crosses borders or provides cross-border services and therefore affects the interests of several Member States. A clear definition of such infrastructure and the entities operating it, and a shared commitment to protect them, is in the interest of all Member States. The Commission encourages Member States to subject critical infrastructure operators to stress tests on the basis of a common set of rules developed at Union level.

3.2 Poland

According to the assumptions of the Crisis Management Act, it is the owners, as well as sole and dependent owners of critical infrastructure elements, who are responsible for the adequate protection of objects, areas, installations and equipment identified as critical infrastructure of the state. It is also their responsibility to develop plans for the protection of the critical infrastructure they own and to maintain their own back-up systems to ensure security and sustain the functioning of the critical infrastructure element until it is restored [4].

Accordingly, responsibility for protecting critical infrastructure lies with its operators. However, given its importance to the State's existence, the Act also assigns infrastructure protection duties to public administration entities. This is natural especially bearing in mind the potential effects of a critical infrastructure failure on people, their property, the national economy or the environment. Consequently, regardless of whether the critical infrastructure is under the management of the public administration or remains entirely under the ownership of economic operators, the expected consequences of a critical infrastructure failure result in critical infrastructure protection tasks being carried out at all levels of the state administration.

In the context of transport, critical infrastructure protection regulations are found in legislation covering various sectors. These include:

- Act of 3 July 2002. Aviation Law - covers content concerning
- air navigation, ground airport infrastructure and aircraft. The issue of the protection of civil aviation against acts of unlawful interference threatening the safety of aviation and persons and property is contained in separate acts and international agreements and regulations.
- Railway Transport Act of 28 March 2003 refers to railway infrastructure, part of which constitutes critical infrastructure. Specifically defined in the Act, railway lines of state significance, railway lines of defence significance and exclusively defence significance are to be considered as an essential part of railway infrastructure.
- The Regulation of the Council of Ministers of 21 April 2022 on objects of particular importance for the security or defence of the state and their special protection, defines objects of particular importance for the security and defence of the state. The comprehensive catalogue, which includes nineteen categories of facilities, includes, inter alia,

formation and service facilities, facilities related to mineral extraction, telecommunication facilities, state reserve warehouses, hydroelectric dams and facilities, power plants and electricity facilities, as well as facilities subordinate to or supervised by the Minister of Defence.

In accordance with the philosophy adopted by GCS, the protection of critical infrastructure should be understood collectively as the protection of:

- physical,
- technical,
- personal,
- ICT,
- legal,
- and plans for its reconstruction.

Physical protection includes: protection of persons, understood as activities aimed at ensuring the safety of life, health and personal integrity, and protection of property, i.e. activities preventing crimes and offenses against property, as well as securing the entry of unauthorized persons to the protected area. It is carried out by security personnel who "physically" defend access to facilities, devices, installations or services of critical infrastructure.

Technical protection is a set of projects related to the construction and operation of facilities, devices, installations and services of critical infrastructure, including technical protection measures aimed at minimizing the risk of disruption in the functioning of CI.

Personal protection is a set of measures and procedures aimed at minimizing the risk resulting from the actions of employees and service providers who, through authorized access to facilities, devices, installations and services of critical infrastructure, may cause disruptions in its operation.

ICT protection is a set of projects and their procedures aimed at minimizing disruptions in the functioning of CI related to the use of this type of infrastructure of ICT systems and networks. This means protection against hacker attacks and cyberterrorism and effective prevention of such incidents.

Legal protection of critical infrastructure is a set of projects aimed at minimizing the risk associated with the activities of other economic, state or private entities whose activities may lead to disruptions in the functioning of CI facilities, devices, installations and services (e.g. hostile takeovers, mergers or sales certain infrastructure elements, which may result in disruptions in its operation).

Activities in the field of critical infrastructure protection carried out at the Polish level are part of the broader European context, as evidenced by the European Critical Infrastructure Protection Program implemented within the European Union.

For activities under the European Critical Infrastructure Protection Program consist of:

- Council Directive 2008/114/EC of 8 December 2008 on the identification and designation of European critical infrastructure and the assessment of the needs to improve its protection,
- instruments financing activities in the field of critical infrastructure protection - in 2007–2013: the "Prevention, preparedness and management of the effects of terrorism

and other types of security risks" program - CIPS, in 2014–2020 the "Internal Security Fund" instrument - ISF,
- activities supporting Member States in implementing the directive (including information exchange system - CIWIN),
- external dimension - the concept of cooperation with third countries in which infrastructure is located in the territory, which in the event of disruption or destruction may affect countries' infrastructure Member States (Council conclusions on the development of the external dimension European Critical Infrastructure Protection Program),
- possible assistance to Member States in working on solutions national in the field of CI.

Critical infrastructure plays a key role in the functioning of the state and the citizens. As a result of events caused by natural forces or as a consequence of human actions, critical infrastructure can be destroyed, damaged and its operation can be disrupted, thereby gambling with the lives and property of citizens. At the same time, such events negatively affect the economic development of the country. Hence, the protection of critical infrastructure is one of the priorities facing every state, including Poland - especially in the geopolitical situation in which we find ourselves after 24 February 2022.

4 Critical Infrastructure in EU Research

EU security research is accompanied by multiple challenges, which require multi-track technological development to overcome, as well as a broad spectrum of management skills. Cybercrime and terrorism are gambling with vulnerable urban or protected areas and critical infrastructure facilities such as ports or transport and energy infrastructure. Development of a package of innovative solutions can help protect these vulnerable areas.

Research efforts at EU level aimed at disaster risk reduction mainly focus on the effective implementation of international conventions, including the UN Sendai Framework for Disaster Risk Reduction, EU regulations such as the EU Civil Protection Mechanism, and policies in the field (e.g. the Floods Directive, the EU CBRN Action Plan, etc.). More broadly, they are intended to foster closer cooperation between the EU and Member States and improve coordination in the field of civil protection, contributing to more effective systems for preventing, preparing for and responding to natural and man-made disasters. This policy framework has a direct impact on how disaster risks are managed in the EU, thus influencing the activities of the emergency services.

Combining these services results in the development of communication systems to facilitate disaster management, linking situational awareness and command centres. All these scenarios concern the operation of emergency services, so EU-funded research teams are working on innovative solutions to support precisely these services in dealing with crisis situations. Finally, the issue of the hazards faced by the civilian population is equally important, so the aim of new technologies and policies is to increase awareness of the risks and thus the resilience of citizens, without forgetting the cultural specificities of different disaster scenarios.

5 Financing of Critical Infrastructure - Transport Sector

Financing the development and maintenance of critical infrastructure is the domain of the state, or, looking more broadly, of public institutions. In accordance with the National Programme for the Protection of Critical Infrastructure [6], critical infrastructure protection activities are financed from the Programme participants' own resources and planned in their budgets:

- in case of administrations, based on Article 26 (1) and (2) of the Crisis Management Act,
- in case of operators of critical infrastructure, based on Article 6 (5) of the Crisis Management Act.

It should be mentioned that in the Programme, activities aimed at development and protection of the transport system were assigned to the minister in charge of transport and the minister in charge of maritime economy. On the other hand, one of the activities of the Council of Ministers under the system of critical infrastructure protection is to enable the obtaining of funds for protection, taking into account these tasks in the state budget.

Most infrastructure projects in the transport sector are at the same time investments in critical infrastructure, although the socio-economic benefits of the solutions developed usually come first. At the national level, this task is carried out by the central administration, nevertheless local government investments are also included, in particular road investments, but also those related to rail transport development.

As previously indicated, the concept of critical infrastructure does not only refer to those elements that are of military and defence importance. In the past, however, it has been one of the most significant factors, often determining the shape of transport networks in a given territory. Examples include railway lines built in the Russian partition in the second half of the 19th century and the limitation of the transport network in the then border region with the Prussian partition, or the construction of a railway line as a bypass for the Warsaw junction in the 1950s[1].

Transport infrastructure requires significant financial resources for its construction and modernisation. For this reason, it constitutes a significant item in the budgets of successive EU cohesion policy programming periods. Within the Structural Funds, the question of whether a given piece of infrastructure constitutes a critical part or is important from the point of view of defence is not the subject of direct intervention. This does not mean, of course, that sections of roads, railways and other transport systems being built or upgraded will not constitute critical infrastructure.

Financing investment projects directly related to defence remains the domain of the state budget - funds for this purpose come from the budgetary part at the disposal of the Minister of National Defence and are disbursed through the managers of the relevant infrastructure systems. Examples of such measures include the reconstruction of railway line no. 410 between Złocieniec and Drawsko.

At the European level, a change in the previous policy of not involving EU funds in the financing of defence-related tasks was initiated by the Connecting Europe Facility

[1] Railway line 12 Skierniewice - Łuków, 160 km long, opened in 1953.

(CEF). One of the objectives identified for the transport component is the adaptation of infrastructure in the TEN-T network to improve civilian and military mobility.

Dedicated calls for military mobility projects (Military Mobility) serve this purpose. Under this component, support can be sought for dual-use investments which, in addition to solving a specific transport problem of economic importance (e.g. removing bottlenecks on the transport network), also offer the possibility of military use (e.g. the possibility of rapid military relocation through the upgrading of a railway line and the construction/restoration of loading sites).

It is worth mentioning that Polish applicants receive funding from it. In 2022, the call under the second competition of the Connecting Europe Facility for Military Mobility, announced by the European Commission in cooperation with the Executive Agency for Climate, Environment and Infrastructure (CINEA), ended. As part of it, Polish beneficiaries submitted a total of eight applications:

- five applications were submitted by airports located in the TEN-T network and concerned airport infrastructure,
- one application related to the construction of loading points on the railway network,
- two applications related to the reconstruction of engineering structures on the national road network.

The total value of the projects is EUR 171 million and the amount of funding requested is over EUR 85 million.

Five Polish projects received funding of over EUR 30 million.

These are:

- The city of Częstochowa project - Maintaining the strategic importance of Wojska Polskiego Avenue in Częstochowa for national defence;
- GDDKiA project - Bridge structures on selected national roads in the TEN-T network located on routes predefined for troop movements;
- PKP PLK project - Works at cargo points located on the TEN-T network;
- GTL S.A. project - Construction of a multimodal goods and fuel delivery hub with railway siding connection at Katowice Airport;
- PPL project - Warsaw Chopin Airport - Investments in dual-use infrastructure to ensure base Service Operations 24/7.

The critical infrastructure challenges for the years ahead are multi-faceted and, as the example of 2020 - 2022 shows, not always and not entirely predictable. Events such as the COVID-19 pandemic and Russia's aggression against Ukraine have shown that today's world is highly unpredictable. The digital side of reality is becoming increasingly important, including in the sphere of transport. Also, Poland's status as a country that is narrowing the development gap to the more developed EU countries, and will therefore receive a smaller pool of funds from the cohesion policy, forces it to look for new solutions in terms of financing infrastructure activities. The coming years must therefore be devoted to well-planned activities both in the sphere of real security of critical infrastructure and strategic planning in the field of financing broadly understood public investments.

6 Summary

The definition of Polish critical infrastructure was introduced in the Crisis Management Act of April 26, 2007. It is specified that this type of objects should be understood as systems and functionally related objects that constitute them, including buildings, devices, installations, services that are key to the security of the state and its citizens and serve to ensure the efficient functioning of public administration bodies, and also institutions and entrepreneurs.

Critical infrastructure in particular may be exposed to various types of threats. There are basically three threats: forces of nature, equipment failures and human actions or omissions.

Is the nearest viaduct, bridge or tunnel a critical infrastructure on which the functioning of the country depends? We don't know that. The list of facilities, railway lines and road networks belonging to critical infrastructure is secret and closely guarded. The reason is obvious and indisputable. Due to the confidentiality of the list, we do not know the scale of incidents related to violations of their security. We know as much as the media tells us. This usually involves unusual, large-scale accidents.

From the point of view of each of us, the nearest stop or platform is critical infrastructure because if it is closed, we will be excluded from transport. Everyone using city public transport is similarly excluded if it is attacked by cybercriminals. Just a decade ago, these were exceptional cases. Today, in the era of hybrid warfare, there are more and more cases. Additionally, global warming causes more and more frequent threats for which we cannot actually prepare. All this means that we should increasingly expand critical infrastructure protection systems. This is possible, among others, by: funds from EU funds.

References

1. Cichoń, B.: Crisis management system in the context of ensuring public safety. In: Kosowski, B., Włodarski, A. (eds.) I International Scientific Conference. Challenges of Civil Security in the XXI Century - Engineering Activities in Science, Research and Practice, p. 28. The Main School of Fire Service, Warsaw (2007)
2. Act of 26 April 2007 on crisis management. J. Laws (2007)
3. Żuber, M.: State critical infrastructure as an area of potential terrorist attack. WSOWL, pp. 181–182. http://www.rocznikbezpieczenstwa.dsw.edu.pl/fileadmin/user_upload/wydawnic-two/RBM/RBM_artykuly/2014_8_11.pdf
4. Tyburska, A.: Critical infrastructure protection overview. Szczytno, p. 151 (2012)
5. EU Council Directive 2008/114/EC of 8 December 2008 on the identification and designation of European Critical Infrastructure and the assessment of the need to improve their protection
6. National Programme for the Protection of Critical Infrastructure (2023). https://www.gov.pl/web/rcb/narodowy-program-ochrony-infrastruktury-krytycznej

Mountain Without Barriers - Method of Evaluating Tourist Trails for Possibilities of Traveling by People with Disabilities Using Specialized Off-Road Vehicle

Marcin Jacek Kłos(✉), Paweł Prusicki, and Kamil Góźdź

Faculty of Transport and Aviation Engineering, Silesian University of Technology, Katowice, Poland

{marcin.j.klos,pawel.prusicki,kamil.gozdz}@polsl.pl

Abstract. The article focuses on route planning in selected mountain areas, which is part of the second component of the project. An analysis of the factors influencing route possibilities to complete the vehicle under development was carried out. Elements relating to transit and passenger comfort are included. In order to reduce the number of tests on the vehicle under development in the environment, a method was developed. A method has been developed to select routes for further tests in real life. Both linear infrastructure (e.g., hiking trails) and point infrastructure (e.g., stopping places and tourist service areas) must meet the adopted assumptions to make travel possible. The research presents a method of inventorying tourist infrastructure to determine the feasibility of travel with the specialized off-road vehicle under development.

Keywords: GIS · spatial analysis · transport accessibility · disabled people · non-urbanized areas · tourist routes

1 Introduction

Transport accessibility is described as one of the basic human needs [1–3]. On the one hand, it makes it possible to satisfy the needs for daily travel (e.g., work, shopping) and optional travel (e.g., trips to nature) [4, 5]. Transport accessibility is measured in various ways in transport; a distinction is made between time accessibility, distance accessibility, and economic accessibility, among others [6, 7]. Accessibility has a huge impact on the development of areas, so efforts should be made to increase it in various fields. This is directly related to the ease of achieving a specific goal [8, 9].

Distinguishing transport accessibility for people with disabilities includes studies on public transport and infrastructural investments to improve accessibility for wheelchair users in towns for compulsory travel [10]. Inadequate infrastructure and lack of access for people with disabilities contribute to social exclusion [11]. The needs of people with disabilities should be taken into account when planning and organizing different spheres of activity. The facilitation of life and the opportunity to enjoy different spheres (e.g.,

G. Sierpiński et al. (Eds.): TSTP 2023, LNNS 907, pp. 114–123, 2024.
https://doi.org/10.1007/978-3-031-53181-1_8

tourism, entertainment) levels the playing field in society. In order to develop in accordance with the policy of sustainable development, access to tourism should be provided to people with disabilities. According to the Polish Tourist and Sightseeing Society PTTK, the concept of tourism for people with disabilities is defined as: "intentional, purposeful physical activity adapted to individual needs, performed in various forms, closely related to sightseeing activity. It is a form of social rehabilitation for people with disabilities. This activity aims to maximize physical, mental, social, and occupational fitness as well as adaptation to normal life. Tourism for people with various disabilities is a recreation and a means of therapy and education" [12]. The definition presented indicates and emphasizes the importance of socializing people with disabilities through tourism.

The development of accessibility of tourism offers for people with disabilities is one of the indicated goals in the government program Accessibility Plus for 2018–2025, developed by the Ministry of Investment and Development. As part of the competition 'Things for people' carried out by the National Center for Research and Development, the Faculty of Transport and Aviation Engineering of the Silesian University of Technology received funding for the implementation of the project entitled 'Integrated platform for planning, organization, supervision and support for the availability of mountain tourism offer for people with difficulties in physical functioning and a specialized off-road vehicle for the implementation of the tourist offer - MOUNTAINS WITHOUT BARRIERS'. The project, in general, is described in the paper [13] by the project manager. Existing solutions currently present a major barrier related to price and, in many cases, only focus on vehicles without additional information systems [14].

The project aims to increase the accessibility of the tourist offer in non-urbanised areas for people with disabilities, including individual needs, due to a lack of full mobility. The project has three main components: the development of a specialized off-road vehicle, an integrated information system for planning, the organization of supervision and support, and the accessibility of mountain tourism for people with special needs. This problem is a very important fact that The European Union population is ageing, and as a result, there will be more people with disabilities. According to the World Health Organization (WHO), there are 1 billion people with disabilities worldwide, representing 15% of the global population [15]. The population in Poland is also ageing, with more than 9 million people over the age of 60.

Tourism for people with disabilities is a partially excluded area. In areas next to the sea, special footbridges and footpaths are being created to overcome the exclusion partially. A major problem remains with mountain areas, which are inherently less accessible. Surveys of people with disabilities indicate that they want to participate in life in all spheres, including holidays [16]. Creating infrastructure to enable people with disabilities to participate in tourism is important in order to overcome social exclusion [14]. The condition of the infrastructure is another important aspect of travel, but there are many effective methods for assessing it [17, 18] and describing the trajectories of vehicle movement [19].

The article focuses on route planning in selected mountain areas, which is part of the second component of the project. An analysis of the factors influencing route possibilities to complete the vehicle under development was carried out. Elements relating to transit and passenger comfort are included. In order to reduce the number of tests on the vehicle under development in the environment, a method was developed. A method has been developed to select routes for further tests in real life. Both linear infrastructure (e.g., hiking trails) and point infrastructure (e.g., stopping places and tourist service areas) must meet the adopted assumptions to make travel possible. The research presents a method of inventorying tourist infrastructure to determine the feasibility of travel with the specialized off-road vehicle under development.

A case study was made based on the limitations of the technical prototype of the vehicle and included other identified conditions which must be fulfilled. The conducted research was carried out in the southern part of the Silesian voivodeship in Poland.

2 Method

In order to identify, using spatial analysis tools, mountain routes that the developed off-road vehicle can potentially operate, the project has developed a method for searching for routes that meet the conditions for making the trip.

The purpose of the developed method is to reduce the necessary testing of the developed off-road vehicle in real conditions and narrow the test trips to routes that people with disabilities can potentially use. The method is the first stage before implementing the routes recommended for rides. Figure 1 shows a scheme of the method.

The first step was to analyze the factors to be taken into account. On the basis of expert knowledge, linear infrastructure (mountain trails), travel start and end points, and intermediate points were taken into account. It was determined that the starting and ending points of the trip must be linked to parking lots that will allow a safe change of means of travel. Intermediate points should be mountain refuges. As part of the analysis of linear infrastructure, inclines were taken into account, which are constraints imposed by the constructed vehicle.

The second step is to identify data sources and collect them. In addition, the quality of the data should be assessed. For example, discontinuities on path lines result in the inability to analyze gradients. Missing data should be obtained from other sources or supplemented manually based on available online sources.

Fulfilment of the indicated specified requirements from the first stage gives recommendations for the given routes to be driven under real conditions and verification of the route for actual driveability without third-party assistance.

The spatial analysis assumes using geographic information system (GIS) data and software (e.g. QGIS [20]) for their processing. The input data for the spatial analysis may come from open data sites (e.g. OpenStreetMap [21]) as well as from the national transport institution or the police databases [22]. The data should be written in the database with coordinates, which allows us to visualize and process them with GIS software.

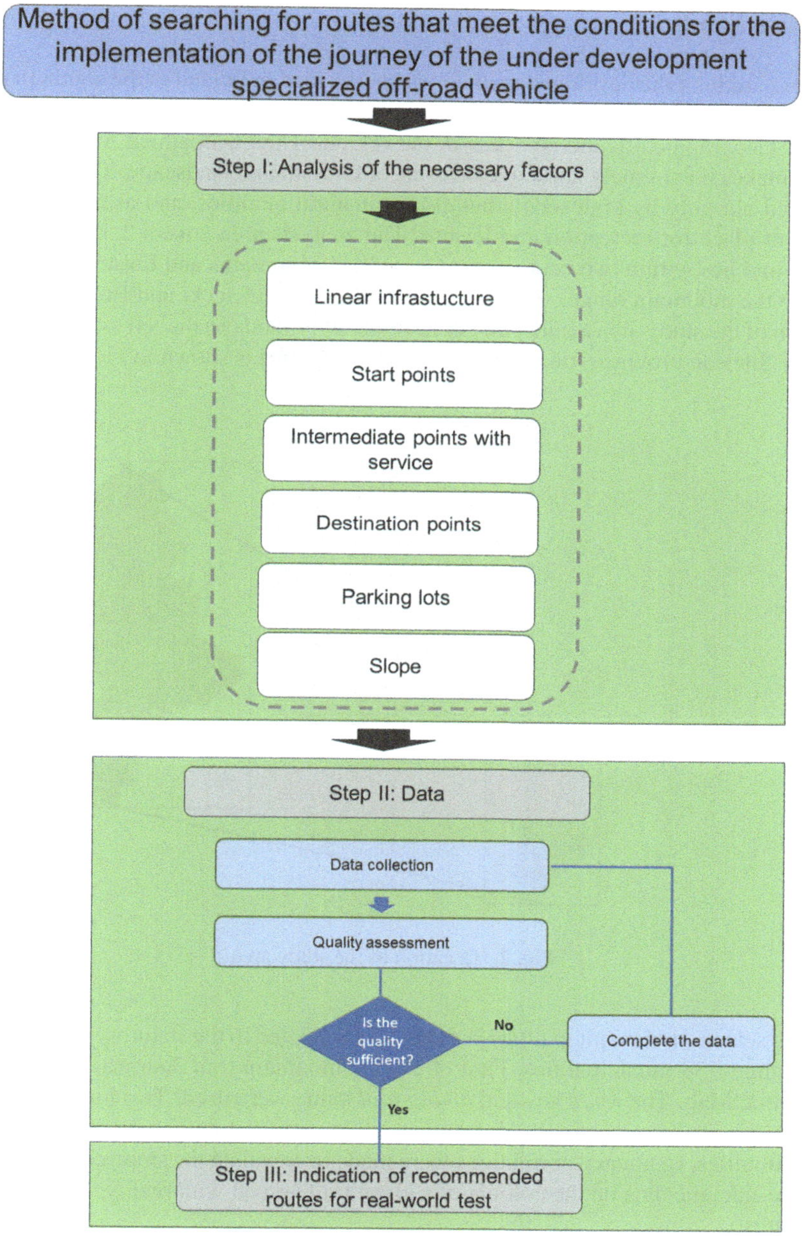

Fig. 1. Scheme of the method of searching for routes that meet the conditions for the implementation of the journey of the under-development specialized off-road vehicle

3 Case Study

The case study of the developed method was performed in Poland in the southern region of the Silesian Voivodeship. The presence of mountains characterizes the indicated region. Geographically, the area lies in the region of the Carpathian Mountain chain, which makes it extremely attractive in terms of tourism and landscape. The area is characterized not only by high scenic but also by natural qualities, and as a result, it is an excellent place for recreation and leisure. The analysis area covers 2,354 square kilometres and lies within two ecoregions: Carpathian Mountains and Eastern Plain, which include the mountain ranges of Beskid Śląski, Beskid Żywiecki and Beskid Mały. The location of the study area against the background of Poland (on the left side of the figure) and the Silesian province (on the right side of the figure) is shown in Fig. 2.

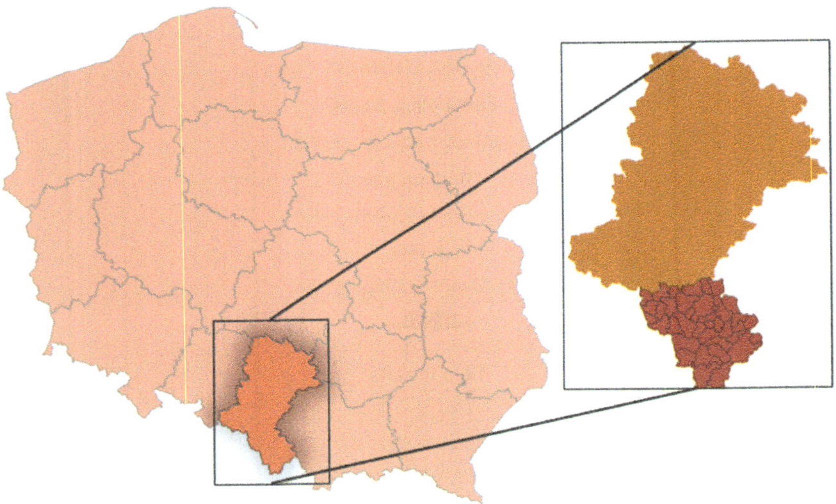

Fig. 2. Location of the study area

The selection of municipalities in the area was related to the statutory functional area consisting of 38 municipalities. First of all, the mountain trail data was checked using OpenStreetMap. The *route* key and *hiking foot* value were used. Two hundred nineteen mountain trails were acquired. Unfortunately, some of the data were characterized by discontinuities, so manual corrections were made to calculate the slope correctly. On the positive side, the data on the colours of a given trail agreed with reality. Figure 3 shows the results.

For 32 of the 38 municipalities, the existence of mountain trails was identified. Subsequently, an analysis was made of intermediate places assumed to be services for people with disabilities to support basic life functions. Due to the required certainty of the functionality of a facility of this type, only mountain huts were taken into account, the functionality of which in Poland is defined by law. Figure 4 shows the analysis of mountain huts. The data was downloaded from OpenStreetMap using the *tourism* key

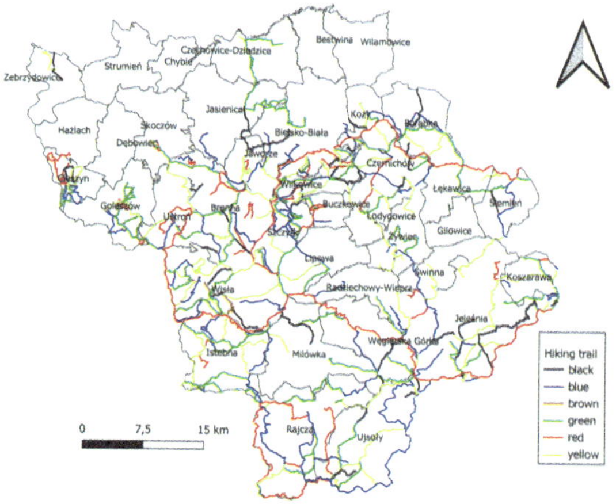

Fig. 3. Linear infrastructure of mountain trails in the analyzed area

and *alpine_hut* value. There are 32 mountain refuges in the area, which serve a total of 97 sections of mountain trails. Trails without mountain huts were eliminated from further analysis.

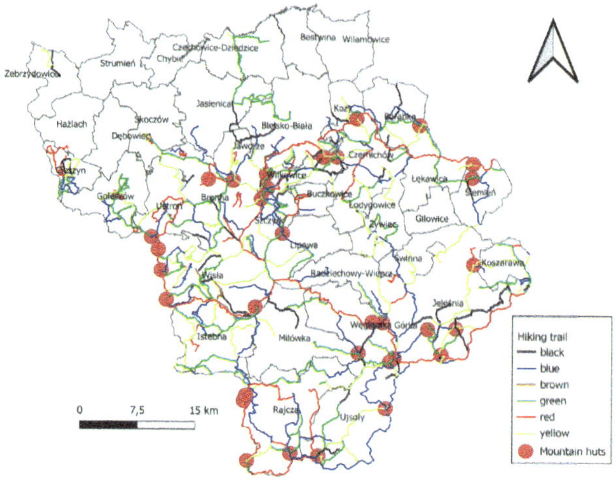

Fig. 4. Point infrastructure of mountain huts in the analyzed area

In the next step, parking spaces were identified, which defined the starting and ending points of the trip. The data was obtained from OpenStreetMap using the *amenity* key with the value of *parking*. Trails away from parking areas were eliminated from further analysis, leaving 80 sections of mountain trails. Figure 5 in part (a) shows a well-connected example with the possibility of leaving the vehicle in Bialy Krzyz and driving the route to Skrzyczne hill, where there is a mountain hostel and a descent to the city centre. On the other hand, part b presents an example of trails without parking infrastructure and mountain hostels.

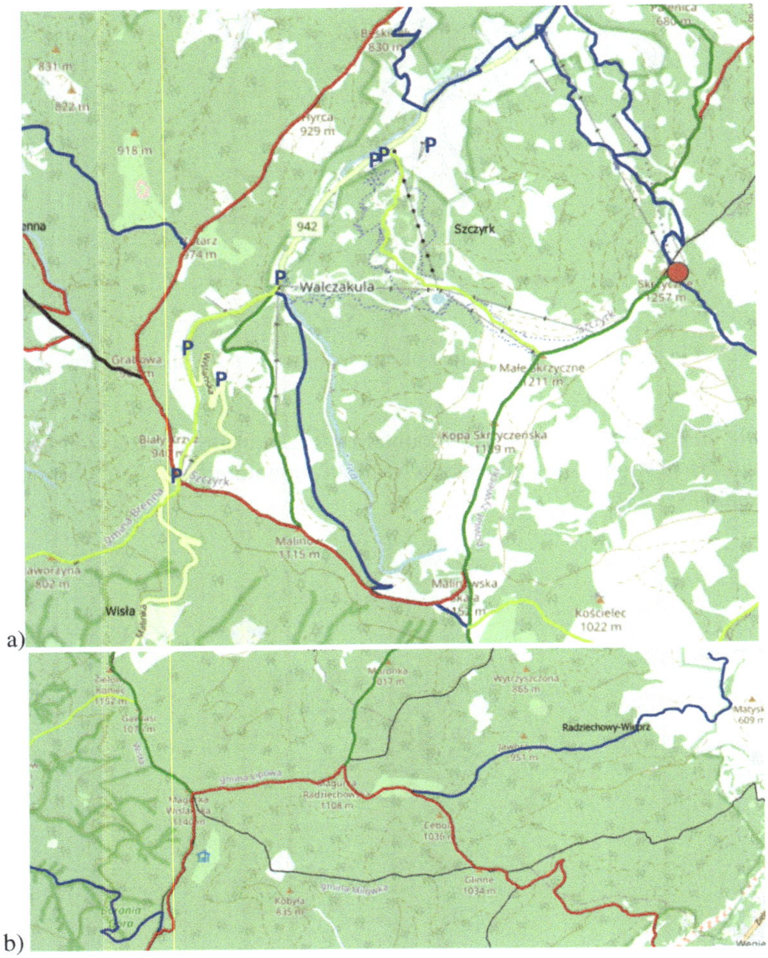

Fig. 5. A) well-connected example with the possibility to leave the vehicle in Bialy Krzyż and take the route to hill Skrzyczne, where there is a mountain hut and downhill to the town centre. B) an example of trails without parking facilities and mountain huts

The last data is slope information. For the vehicle under development, the boundary parameters of going uphill 15% and driving downhill 20% were determined. Figure 6 shows one case. For the green lines, the going uphill and driving downhill conditions are met, while the yellow colour indicates limits in the range of 13% to 17%. The red colour, on the other hand, indicates overshoot. The data obtained for calculating the gradients came from the Numerical land cover model from the National Geoportal.

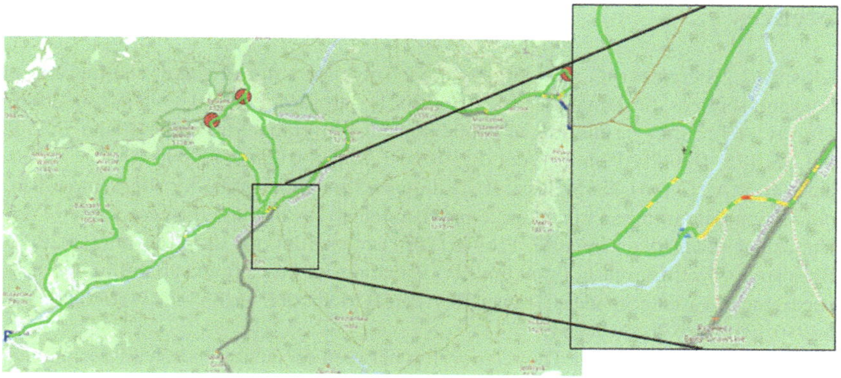

Fig. 6. Slope analysis

The indicated route, starting with a parking lot in Zlatna Huta was recommended for testing in real conditions. The tests indicated that the problematic 4-m section is manageable in good conditions, as mud can prevent the section from being traversed.

4 Conclusion

The developed method makes it possible to perform analyses of large data sets and supports decision-making for testing the vehicle under development in real conditions. The biggest difficulty is to perform data quality assessment. Linear data mountain trails require continuity and accurate localization, as this directly affects the quality of elevation calculations. The development of a database of trails that allow the developed vehicle to drive, along with the subsequent test, will allow people with disabilities to perform the drive independently without problems. It is important to remember that in addition to the input data, other obstacles may not be detected using spatial data. Examples include tree felling and the effect of weather conditions on surface quality.

Acknowledgements. The present research has been financed through the National Centre for Research and Development as a part of the competition within the scope of "Things are for people" in project with title: "Integrated platform for planning, organization, supervision and support for the availability of mountain tourism offer for people with difficulties in physical functioning and a specialized off-road vehicle for the implementation of the tourism offer - Mountains without barriers" realized by Silesian University of Technology.

Rzeczpospolita Polska   Dostępność Plus   NCBR   Politechnika Śląska

References

1. Kłos, M.J., Sierpiński, G.: Building a model of integration of urban sharing and public transport services. Sustainability **13**, 3086 (2021). https://doi.org/10.3390/su13063086
2. Murray, A.T.: Strategic analysis of public transport coverage. Socioecon. Plann. Sci.. Plann. Sci. **35**, 175–188 (2001). https://doi.org/10.1016/S0038-0121(01)00004-0
3. Kolcsár, R.A., Csikós, N., Szilassi, P.: Testing the limitations of buffer zones and urban atlas population data in urban green space provision analyses through the case study of Szeged, Hungary. Urban Urban Green **57**, 126942 (2021). https://doi.org/10.1016/j.ufug.2020.126942
4. Börjesson, M., Eliasson, J., Franklin, J.P.: Valuations of travel time variability in scheduling versus mean-variance models. Transp. Res. Part B Methodol. **46**, 855–873 (2012). https://doi.org/10.1016/j.trb.2012.02.004
5. Jedynak, T., Wąsowicz, K.: The relationship between efficiency and quality of municipally owned corporations: evidence from local public transport and waste management in Poland, pp. 1–30 (2021)
6. Ziari, H., Keymanesh, M.R., Khabiri, M.M.: Locating stations of public transportation vehicles for improving transit accessibility. Transport **22**, 99–104 (2007). https://doi.org/10.1080/16484142.2007.9638106
7. Panagiotopoulos, G., Kaliampakos, D.: Location quotient-based travel costs for determining accessibility changes. J. Transp. Geogr.Geogr. **91**, 102951 (2021). https://doi.org/10.1016/j.jtrangeo.2021.102951
8. Kaszczyszyn, P., Sypion-Dutkowska, N.: Walking access to public transportation stops for city residents. A comparison of methods. Sustainability **11**, 3758 (2019). https://doi.org/10.3390/su11143758
9. Żochowska, R., Kłos, M.J., Soczówka, P., Pilch, M.: Assessment of accessibility of public transport by using temporal and spatial analysis. Sustainability **14**, 16127 (2022). https://doi.org/10.3390/su142316127
10. Soczówka, P., Kłos, M.J., Żochowska, R., Sobota, A.: An analysis of the influence of travel time on access time in public transport, **108**, 133–140 (2020). https://doi.org/10.20858/sjsutst.2021.111.12
11. Oh, J.-S., Smith, C.S., Qatra, R., et al.: Estimating and enhancing public transit accessibility for people with mobility limitations. Transp. Res. Cent. Livable Communities 29p. (2017)
12. Śledzińska, J.: Turystyka osób niepełnosprawnych w Polskim Towarzystwie Turystyczno-Krajoznawczym. Niepełnosprawność–zagadnienia. Probl. Rozw **2**, 81–103 (2012)
13. Staniek, M.: Support system of mountain travels for people. Sci. Pap. Silesian Univ. Technol. Organ. Manag. Nauk Politech Sl Ser Organ i Zarz, 12 (2023)
14. Zabłocki, M., Branowski, B., Kurczewski, P., et al.: Designing innovative assistive technology devices for tourism. Int. J. Environ. Res. Publ. Health **19**, 14186 (2022). https://doi.org/10.3390/ijerph192114186
15. World Health Organization Bank: World report on disability 2011. Summary also available in Braille (2011)
16. Özcan, E., Topcu, Z.G., Arasli, H.: Determinants of travel participation and experiences of wheelchair users traveling to the bodrum region: a qualitative study. Int. J. Environ. Res. Publ. Health **18**, 1–28 (2021). https://doi.org/10.3390/ijerph18052218
17. Staniek, M.: Detection of cracks in asphalt pavement during road inspection processes. Sci. J. Silesian Univ. Technol. Ser. Transp. **96**, 175–183 (2017). https://doi.org/10.20858/sjsutst.2017.96.16
18. Staniek, M.: Road pavement condition as a determinant of travelling comfort. In: Sierpiński, G. (ed.) Intelligent Transport Systems and Travel Behaviour. AISC, vol. 505, pp. 99–107. Springer, Cham (2017). https://doi.org/10.1007/978-3-319-43991-4_9

19. Piecha, J., Staniek, M.: The context-sensitive grammar for vehicle movement description. In: Bolc, L., Tadeusiewicz, R., Chmielewski, L.J., Wojciechowski, K. (eds.) ICCVG 2010. LNCS, vol. 6375, pp. 193–202. Springer, Heidelberg (2010). https://doi.org/10.1007/978-3-642-15907-7_24
20. QGIS Development: QGIS Geographic Information System (2023)
21. OpenStreetMap contributors (2023). Planet dump retrieved from https://planet.osm.org
22. Sierpiński, G., Staniek, M., Kłos, M.J.: Decision making support for local authorities choosing the method for siting of in-city ev charging stations. Energies **13**, 4682 (2020). https://doi.org/10.3390/en13184682

Author Index

G. Sierpiński et al. (Eds.): TSTP 2023, LNNS 907, p. 125, 2024.
https://doi.org/10.1007/978-3-031-53181-1

SPRINGER NATURE

GPSR Compliance

The European Union's (EU) General Product Safety Regulation (GPSR) is a set of rules that requires consumer products to be safe and our obligations to ensure this.

If you have any concerns about our products, you can contact us on ProductSafety@springernature.com

In case Publisher is established outside the EU, the EU authorized representative is:

Springer Nature Customer Service Center GmbH
Europaplatz 3
69115 Heidelberg, Germany

The manufacturer's authorised representative in the EU is Springer Nature Customer Service Centre GmbH, Europaplatz 3, 69115 Heidelberg, Germany. If you have any concerns regarding our products, please contact ProductSafety@springernature.com

Printed and bound by CPI Group (UK) Ltd, Croydon, CR0 4YY
01/10/2026
02226300-0005